A Study Book for the
NEBOSH Certificate in Fire Safety
and Risk Management

RMS Publishing Limited

Victoria House, Lower High Street, Stourbridge DY8 1TA

© ACT Associates Limited.

First Published April 2006.

Cover design by Graham Scriven.

Printed and bound in Great Britain by Antony Rowe Ltd, Chippenham, Wiltshire.

ISBN 1 900420 76 7

Foreword

EDITOR: PATRICK G. COX

With the adoption of risk based fire safety solutions under the Fire Precautions (Workplace) Regulations in 1997, the onus for providing a work environment safe from fire, moved away from the Fire and Rescue Service Inspector, to the owner/occupier/manager of premises used as places of work. This has been further developed under the Regulatory Reform (Fire Safety) Order 2005 which repeals both the Fire Precautions Act 1971 and the Fire Precautions (Workplace) Regulations 1997 and 1999 alongside amending or repealing a host of other legislation covering aspects of fire safety for employees and others in the workplace.

This change presents both an opportunity for employers, employees, the responsible person and the Fire and Rescue Services - and a challenge. To be effective the process must be seen as dynamic - fire risks change all the time in any occupied building - and the solutions and systems put in place as a result of the initial risk assessment, must be constantly reviewed and monitored to ensure they remain appropriate and effective. Risk based solutions for fire safety also depend, for their effectiveness, on every employee understanding the rationale behind certain measures and appreciating the consequences of any lapse or failure. This means that there is an ongoing need for training and commitment on the part of everyone - including the Chairman of the Board of Directors.

This study guide provides an effective and very useful introduction to this important process. Naturally it cannot provide all the answers to every question and some situations will be encountered that will require further research, advice creative thinking to find a practical, workable solution that still allows people to work efficiently and safely without being overly burdensome or expensive. Remember, the best solution is always one which achieves the objective without being unduly obstructive and which is "owned" by the end user - the person you are trying to protect. It is a frequently encountered mistake to find that a solution has been decided and imposed without consulting the people who must work with it. They don't "own" it as a result and will frequently "amend" it, sometimes with serious consequences, often completely defeating the intended purpose..

After many years both as a fire safety inspector and as a trainer of inspectors, I welcome this move away from "prescription" as I believe that, properly applied and managed, it can provide "best fit" solutions for the employers and the employees. It is now many years since fire safety became a legal duty for employers and in that time we have seen huge improvements in building management, construction methods and design incorporating safety principles. Social attitudes and expectations have moved on and the time is now right for the building users, employers and employees to take responsibility for their own safety from the ravages of fire. The NEBOSH Certificate in Fire Safety and Risk Management will, I believe, help to protect lives, property and the wider economy for the future by producing people in industry and commerce who have a sound and practical approach to reducing the risks posed by fire in every walk of life.

Patrick G. Cox, BSc, MA, MIFire E, ASAESI, AIOSH, CFEI, CFII.
Senior Tutor, The Fire Service College, Community Safety Support Centre.

Editor's Notes

Diagrams and photographs

We have taken particular care to support the text with a significant number of photographs. They are illustrative of both good and bad working practices and should always be considered in context with supporting text. I am sure that students will find this a useful aid when trying to relate their background and experience to the broad based NEBOSH Certificate in Fire Safety and Risk Management syllabus. They will give an insight into some of the technical areas of the syllabus that people have difficulty relating to when they do not have a strong technical background.

Where diagrams/text extracts are known to be drawn from other publications, a clear source reference is shown and RMS/ACT wish to emphasise that reproduction of such diagrams/text extracts within the Study Book is for educational purposes only and the original copyright has not been infringed.

Legal requirements

The Study Book has at its heart the fact that health and safety should be managed as a risk. However as one of the risks to a business is the risk of prosecution care has been taken to relate topics to current legislation. The level of treatment is targeted to cover the interests of the Certificate in Fire Safety and Risk Management student. Legislation is referred to in context in the various elements that comprise the Study Book and reflects the syllabus of the NEBOSH Certificate in Fire Safety and Risk Management. In addition the book has an element dedicated to legislation, Element 7, and here the student will find a useful summary of the legislation required by the NEBOSH Certificate in Fire Safety and Risk Management.

The NEBOSH General Certificate does not examine the student's knowledge of Section numbers of the Health and Safety at Work Act or knowledge of Regulation numbers. These are referred to in the Study Book in order to differentiate different components of the law and to aid the student in referencing legislation in the workplace, if required by their work.

Syllabus

Each elements of the Study Book has a element overview that sets out the aims of the elements, the content and learning outcomes. The Study Book reflects the order and content of the NEBOSH Certificate in Fire Safety and Risk Management syllabus and in this way the student can be confident that the Study Book reflects the themes of the syllabus. In addition, the syllabus is structured in a very useful way, focusing on hazards, their control and core management of health and safety principles.

National Vocational Qualification

We are confident that those working towards national vocational qualifications will find this Study Book a useful companion for NVQ Level 3 in Occupational Health and Safety.

Acknowledgements

Managing Editor: Ian Coombes MIOSH - Managing Director, ACT; NEBOSH Advisory Committee and Board member; IOSH Professional Affairs Committee member.

RMS Publishing and ACT Associates Ltd wish to acknowledge the following contributors and thank them for their assistance in the preparation of the Certificate in Fire Safety and Risk Management study book: Roger Chance, Jim Clifford, Dean Johnson, Barrie Newell, Clive Raybould, Julie Skett and Luane Steele.

NEBOSH Study Books also available from RMS:

NEBOSH National General Certificate (3rd Edition)   ISBN 1 900420 68 6

NEBOSH National Certificate in Construction (1st Edition)   ISBN 1 900420 58 9

NEBOSH Level 4 Diploma (1st Edition):
A. Managing Health and Safety   ISBN 1 900420 59 7
B. Hazardous Agents in the Workplace   ISBN 1 900420 60 0
C. Workplace and Work Equipment   ISBN 1 900420 61 9

Contents

Figure List (including tables and quotes)

List of abbreviations

LEGISLATION

CHIP	Chemicals (Hazard Information and Packaging for Supply) Regulations 2002
COMAH	Control of Major Accident Hazards Regulations 1999
CSR	Confined Spaces Regulations 1997
DDA	Disability Discrimination Act 1995
DSEAR	Dangerous Substances and Explosive Atmospheres Regulations 2002
FAR	Health and Safety (First Aid) Regulations 1981
FRA	Fire and Rescue Services Act 2004
HASAWA	Health and Safety at Work etc Act 1974
HSCER	Health and Safety (Consultation with Employees) Regulations 1996
IER	Health and Safety Information for Employees Regulations 1989
MHSWR	Management of Health and Safety at Work Regulations 1999
PPER	Personal Protective Equipment at Work Regulations 1992
RIDDOR	Reporting of Injuries, Diseases and Dangerous Occurrences Regulations 1995
RRFSO	Regulatory Reform (Fire Safety) Order 2005
SRSC	Safety Representatives and Safety Committees Regulations 1977
SSSR	Health and Safety (Safety Signs and Signals) Regulations 1996
WRA	Water Resources Act 1991

GENERAL

ABI	Association of British Insurers
ABS	Acrylonitrile Butadiene-Styrene
ACOP	Approved Code Of Practice
BS	British Standards
CCNSG	Client Contractor National Safety Group
CO_2	Carbon Dioxide
CO	Carbon Monoxide
dB	Decibel
EA	Environment Agency
ECITB	Engineering Construction Industry Training Board
EEA	European Economic Area
EMAS	Employment Medical Advisory Service
EU	European Union
FD	Fire Door
GEEP	Generic Emergency Evacuation Plan
H	Hydrogen
HSC	Health and Safety Commission
HSE	Health and Safety Executive
IT	Information Technology
LEL	Lower Explosive Limit
LPG	Liquefied Petroleum Gas
O	Oxygen
ODPM	Office of the Deputy Prime Minister
OH	Hydroxyl radicals
PAS	Publicly Available Specification
PEEP	Personal Emergency Evacuation Plan
PPE	Personal Protective Equipment
PTFE	Polytetraflouroethylene
PVC	Polyvinyl Chloride
RES	Representatives of Employee Safety
RMFAS	Remotely Manned Fire Alarm System
RX	Receiver
TUC	Trades Union Congress
TX	Transmitter
UEL	Upper Explosive Limit

Managing Fire Safety

Overall Aims

On completion of this Element, candidates will understand:

- reasons for managing fire risks.
- the legal framework relating to fire and health and safety.
- a framework for managing fire safety.
- the legal and financial consequences of failure to manage fire safety.
- the process, sequences and the purpose of investigating fire related incidents (accidental and maliciously set fires) in the workplace.
- the legal and organisational requirements for recording and reporting fire related incidents.

Content

Specific Intended Learning Outcomes

The intended learning outcomes of this Element are that candidates will be able to:

1.1   outline the legal framework for the regulation of fire safety

1.2   describe the roles and powers of enforcement agencies and the judiciary

1.3   outline the key elements of a fire safety management system

1.4   explain the purpose of and the procedures for investigating fires in the workplace

1.5   describe the requirements for the statutory reporting of fire-related fatalities and injuries in the workplace

1.6   liaise with specialists and other investigators.

Relevant Statutory Provisions

The Regulatory Reform (Fire Safety) Order (RRFSO) 2005

1.1 - Maintaining good standards of fire safety

General argument

There are three good reasons for preventing fires in the workplace and maintaining good standards of fire safety.

1. Moral

Fires result in a great deal of suffering for those affected. We must all do what we can to avoid this.

2. Legal argument

It is a legal requirement to prevent fire, protect employees and other relevant persons from the effects of fire and to mitigate the effects of fire from anyone in the vicinity of a premises on fire.

3. Financial argument

The costs to society as a result of fire are substantial, especially when we add on the consequential losses such as interruption to business and the loss of trade. The environmental damage caused by fires can also be enormous. It should be noted that most companies that have a major fire never resume business again.

The size of the fire safety 'problem'

NUMBERS OF FIRE FATALITIES AND INJURIES

The following figures are based on the information in the 2004 UK fire statistics which are published by the Office of the Deputy Prime Minister (ODPM).

The fire and rescue services attended almost 900,000 fires and false alarms. This represents an 18% decrease on the 2003 figures. Within this figure, fires decreased by 29% to 442,700, while false alarms fell by 5% to 448,800.

Total number of fires attended   -   442,700.

Fires attended in dwellings   -   59,700 (6% reduction on 2003).

The number of building fires   -   97,000.

Total number of *accidental* primary fires 97,700.

- 47,800 fires (49%) were in dwellings.
- 21,500 (22%) in other buildings.
- 17,800 (18%) in road vehicles.
- 10,600 (11%) in outdoor structures / secondary fires with casualties or 5 or more appliances.

The key changes in the 2004 statistics compared to those for 2003 were:

Accidental primary fires	2003	2004	Difference
Dwellings	50,000.	47,800	- 4%
Other buildings	23,100.	21,400	- 7%
Road vehicles	19,900	13,300	- 33%

Figure 1-1: Accidental primary fires. *Source: ODPM 2004.*

The number of *deliberate* primary fires fell for the third consecutive year, by 21% from 115,100 in 2003 to 91,200. The main reason for the fall in 2004 was a 25% reduction in deliberate fires in road vehicles.

Casualties from fires were:

- Fire related deaths - 508 or down by 14% on 2003.
- Non-fatal casualties (including fire fighters) - 14,600 or down by 6% on 2003.

Primary fires include all fires in buildings, vehicles and outdoor structures or any fire involving casualties, rescues, or fires attended by five or more appliances.

Secondary fires are the majority of outdoor fires including grassland and refuse fires unless they involve casualties or rescues, property loss or five or more appliances attend. They include fires in single derelict buildings.

Chimney fires are any fire in an occupied building where the fire was confined within the chimney structure (and did not involve casualties or rescues or attendance by five or more appliances).

A *false alarm* is defined as an event in which the fire and rescue service believes they are called to a reportable fire and then find there is no such incident.

Commercial fire losses remain substantial - £672m in 2003. Fires in commercial buildings remain a serious cost to the UK economy, with insured losses (£672m in 2003) over 50% higher than for domestic fires. In practice, despite the large insured losses arising from commercial fires, actual loss of life is relatively rare - of 540 deaths in fires in 2002, 29 occurred in commercial buildings. Even non-fatal casualties, more common than fatalities, are comparatively rare in commercial fires (45 per 1,000 fires), reflecting the success of the current arrangements. It is this success in reducing fatalities and injuries in workplace fires that is leading to a shift in resources towards domestic fire prevention work, where the majority of casualties occur. Whilst Association of British Insurers (ABI) fully supports a risk-based approach to deploying fire service resources, a wider view of the risks facing society must be taken. The economic and social consequences of fire also include business failures, job losses and the loss of local services and facilities. The insured losses from commercial fires can be split between the direct financial losses caused by fire damage (£672m in 2003) and the indirect cost through lost business (£81m in 2003). These costs vary significantly year on year since fires in substantial commercial enterprises can represent a large proportion of the total losses. Business interruption losses consequent on the fire damage can be several multiples of the physical damage to assets in such cases. Overall Government estimates are that fire costs the UK economy around £6.6bn per year (including both domestic and commercial fires). Economic losses are over six times higher than insured losses.

(Quoted from Association of British Insurers (ABI).

COSTS OF INADEQUATE FIRE RISK MANAGEMENT

- In 2003, the total cost of fire is estimated at £7.7bn, equivalent to approximately 0.9% of the gross value added of the economy.
- The costs as a consequence of fire, including property losses, human casualties and business disruption, are estimated at £3.3bn in 2003.
- The cost to the Fire and Rescue Service (FRS) of providing fire cover is estimated at £1.7bn in 2003. The cost of FRS attendance at both false alarms and non-building fires is estimated at almost £700m respectively, due to the large proportion of all incidents that these account for.
- The total cost of arson in 2003 is estimated at £2.8bn, which includes an allocation of the total costs incurred in anticipation of fire. The cost of FRS response to malicious false alarms is estimated at £83m.
- The average cost of a domestic fire is estimated at £25,000, of which approximately £15,000 is accounted for by the economic cost of injuries and fatalities.
- The average cost of fire in a commercial building is estimated at £58,000, of which the cost of fire damage to property represents £45,000. The average cost of a vehicle fire is estimated at £4,700.

Whilst caution should be exercised in analysing trends over a short period of time, particularly since it is difficult to observe costs directly, the cost of fire remained stable as a proportion of the economy over the last 5 years, at approximately 0.9%.

As can be seen, the true costs of a fire are extremely high. This can be highlighted by looking at two recent examples.

Primark warehouse fire, Leicestershire

Dozens of fire-fighters tackled the massive blaze at the 440,000 sq ft Primark warehouse at Magna Park, near Lutterworth. The warehouse building itself cost £8million to build, with up to an estimated £50million worth of garments being destroyed. The building was owned and operated by TNT on behalf of Primark.

The building was fully insured both for stock loss and business interruption. The impact on the companies has been survivable, but the cost will of course go to the insurers.

Figure 1-2: Primark warehouse fire.   *Source: BBC News.*   Figure 1-3: Buncefield Oil Terminal.   *Source: Royal Chiltern Air.*

Buncefield Oil Terminal fire, Hertfordshire

The Buncefield fuel depot fire in December 2005 was the biggest in the UK's peacetime history. Explosions and heat from the fire caused severe damage to more than 80 buildings on the industrial estates surrounding the terminal, and some were demolished by the blasts. Initial estimates from Hertfordshire Chamber of Commerce put the cost of the damage at between £500 million and £1 billion. Information technology (IT) software and services firm Northgate Information Solutions was the closest business to the fire, when a blast ripped through the oil depot in the early hours of Sunday, 11 December. The explosion rolled across Northgate's car park and into its 150,000 square-foot building, starting a fire which left the firm's UK headquarters an empty shell. This fire was an extreme example with many buildings being destroyed or damaged by its effects.

With the Regulatory Reform (Fire Safety) Order (RRFSO) 2006 requiring that we 'mitigate the effects of a fire on anyone in the premises and anyone in the vicinity of the premises, who may be effected by a fire on the premises', such incidents will create test cases for how this new legislation is applied.

1.2 - The tort of negligence

General points

Common law establishes a duty to take reasonable care of those that might be affected by how people do things. If insufficient care is taken while doing things it may be considered that someone had been negligent. Negligence is a civil wrong, a tort, recognised by civil courts, which means someone affected by the negligence can sue for their loss.

"You must take reasonable care to avoid acts or omissions which you can reasonably foresee would be likely to injure your neighbour".

Figure 1-4: Definition of negligence.   *Source: Donoghue v Stevenson 1932.*

The common law duty of care, and the possibility of being sued for negligence, applies widely to things that are done in daily life, including work situations. A person in control of a building, such as an occupier of the building or an employer, would have a common law duty to take care to prevent harm from fire. If a fire occurs, because insufficient care was taken, then negligence may be considered to have occurred and those harmed by the negligence may sue for their loss. They would sue those that caused the negligence; this may be several classes of person - the owner of the building, the occupier and perhaps a contractor - all may be negligent and may be separately sued. The people harmed, and therefore suing for their loss, may be employees of the occupier/employer, another person's employees, visitors to the building, members of the public or even fire fighters.

TESTS OF PROOF

In order to prove negligence and obtain damages (usually in the form of financial compensation) the injured party must show that:

- They were owed a common law duty of care.
- There was a failure to fulfil the duty to take reasonable care.
- Damage, loss or injury resulted from the breach.

The employer's common law duty of care to employees

In common law the employer must take reasonable care to protect employees from the employer's acts or omissions that may reasonably be foreseen would injure employees. The minimum elements of the common law duty which the employer should pay heed to were summed up in Wilson and Clyde Coal Co Ltd v English (1938). This is a notable case that set a precedent and confirmed what an employer should ensure when meeting common law duties:

- A safe place of work.
- Safe appliances and equipment.
- A safe system of work.
- Competent and safety conscious personnel.

Failure to meet the duty would usually indicate negligence on the part of the employer. If an employee is injured at work as a result of a fire or failure of fire safety equipment they may be able to sue the employer for the tort (civil wrong) of negligence.

The duty of care owed by the occupier of a building

Occupiers of premises have a common law duty of care to take reasonable care for those that might foreseeably be affected by their actions or omissions. This will include consideration of those people that are the occupier's employees and any lawful visitors to the building, for example, members of the public invited to buy goods in a shop, people delivering goods into a warehouse operation or contractors doing work within or on the building. This duty of care will involve consideration of how these people may be affected by a fire in the building. Failure to take reasonable care of these people could lead to harm to them and an action against the occupier for common law tort of negligence. The occupier may not be the owner of the building and the things that they may reasonably do could be limited by this. The owner of the building would also carry a common law duty of care towards the occupier, their employees and lawful visitors to the building and the occupier may be sued by each separately.

It is reasonable to expect that in relation to multi-occupancy premises the various occupiers meet their common law duty to take reasonable care as individuals, and that they co-operate / co-ordinate with all other occupiers to assist them in meeting their duty. If the occupiers are not also the owner of the premises the owner would also need to co-operate and co-ordinate with the occupiers to meet their duty. In practical terms this may include taking part in evacuation drills and allowing the test of fire alarms.

1.3 - The legal framework for regulating fire safety

Main features

REGULATORY REFORM (FIRE SAFETY) ORDER (RRFSO) 2005

As the name suggests this legislation reforms a number of pieces of legislation relating to fire. The main pieces of legislation replaced with this single order are the Fire Precautions Act (FPA), Fire Precautions (Workplace) Regulations (FPWR), Licensing Act and Housing Act. The Regulatory Reform (Fire Safety) Order (RRFSO) 2005 becomes the primary piece of legislation dealing with fire. The legislation applies to most non-domestic premises other than offshore installations, ships, agricultural or forestry land, mines and boreholes. This includes premises relating to the voluntary sector and self-employed in England and Wales (Scotland and Northern Ireland come under separate legislation). For example:

- Factories and warehouses.
- Offices and shops.
- Pubs, clubs and restaurants.
- Hotels and hostels.
- Premises that provide care.
- Community halls.
- Schools.
- Tents and marquees.
- Open air public gatherings (concerts, shows etc).

The RRFSO represents a move towards greater emphasis on fire prevention, through the implementation of measures derived from risk assessments. The RRFSO abolishes Fire Certificates, removing their legal status. However, the fire precautions they imposed cannot be discarded without due consideration and good reason. It replaces fire certification under the Fire Precautions Act 1971 with a general duty to ensure, so far as is reasonably practicable, the safety of employees with regard to fire. In addition, it places a general duty in relation to non-employees to take such fire precautions as may reasonably be required in the circumstances to ensure that premises are safe and also a duty to carry out a risk assessment. The main duty-holder is the "responsible person" in relation to the premises. The duties on the responsible person are extended to any person who has, to any extent, control of the premises. In summary the RRFSO requires the responsible person to:

- Carry out or nominate someone to carry out a fire risk assessment identifying the risks and hazards.
- Consider who may be especially at risk.
- Eliminate or reduce the risk from fire as far as is reasonably practical and provide general fire precautions to deal with any residual risk.
- Take additional measures to ensure fire safety where flammable or explosive materials are used or stored.
- Create a plan to deal with any emergency and, in most cases, document the findings.
- Review the findings as necessary.

The RRFSO becomes the principal piece of legislation regulating fire at work. The RRFSO comprises of 5 parts:

- Part 1   General.
- Part 2   Fire Safety Duties.
- Part 3   Enforcement.
- Part 4   Offences and appeals.
- Part 5   Miscellaneous.

Each part is then subdivided into the individual points or articles as they are called in the order. The RRFSO makes provision for the creation of regulations and is supported by a series of guidance documents in the form of fire safety guides.

See also - Relevant Statutory Provisions - Element 7.

In addition, there is a schedule that covers the risk assessment process, plus details of the various legislation that are repealed or amended.

This RRFSO is very comprehensive, encompassing all of the **general requirements** for fire safety in premises into one document. The onus of responsibility is moved totally to the 'responsible person' to ascertain the control measures that are needed. The requirement to conduct fire risk assessments originally established by the Management of Health and Safety at Work Regulations (MHSWR) 1999 means that existing premises should have risk assessments associated with them that would satisfy requirements of RRFSO. Risk assessments should also be carried out prior to alteration or design of new premises.

FIRE SAFETY REGULATIONS

The RRFSO provides for the Secretary of State to make regulations in relation to risk to relevant persons as regards premises. For example, regulations may be made "for securing that persons employed to work in the premises receive appropriate instruction or training in what to do in case of fire". Regulations made under RRFSO provide **specific legal requirements** to define and regulate fire precautions covered by the general requirements of the RRFSO. It should be noted that regulations carry the full force of the law and breaches can lead to significant prosecutions, just as they might for breaching an Act or Articles within an Order.

There are a number of specific regulations that relate to fire safety. These include regulations such as the Health and Safety (Safety Signs and Signals) Regulations (SSSR) 1996 and the Dangerous Substances and Explosive Atmosphere Regulations (DSEAR) 2002.

APPROVED CODES OF PRACTICE

There is no general government code of practice issued for fire safety. The nearest to this would be the Approved Document B which applies to the construction of new buildings and building alterations. As an approved document it is intended to provide guidance for some of the more common building situations. There is no obligation to follow the solutions contained within the document and people are free to meet the relevant requirements of the Building Regulations in some other way. However there is much fire safety guidance both past and present ('Fire Safety an Employers Guide' - HSE Books as an example). There is also an extensive range of British Standards (BS) that relate to fire safety standards in buildings, for example, the BS 5588 and BS 9999 series all relate to fire safety. In addition the PAS 79, a Publicly Available Specification (PAS), gives a structured approach and corresponding documentation for conducting and recording significant findings of fire risk assessments. Guidance books have been specifically written with the RRFSO in mind as detailed below.

OFFICIAL GUIDANCE

Guidance is provided to support the duties expressed in the RRFSO and suggest ways in which they may be met. They seek to consider the practical application of these duties and as such the Guidance Books for RRFSO are written to be premises type specific. They should all follow a similar format but have a premises specific section when it looks at means of escape and other issues. They are purely for guidance and are aimed at existing premises and therefore should not be used to design fire safety in new buildings, as the building may be subject to Building Regulations. The guides do not set prescriptive standards, but provide recommendations and guidance for use when assessing the adequacy of fire precautions in premises. There is no obligation to adopt any particular solution that is denoted in the guides, however they provide useful information that indicates what may be suitable for the particular type of premise. Fire safety arrangements do not have to be the same as those shown in the guidance, but the responsible person must demonstrate that they meet an equivalent standard. There are 11 guides relating to the RRFSO:

- Fire safety in offices and shops.
- Fire safety in factories and warehouses.
- Fire safety in premises providing sleeping accommodation.
- Fire safety in premises providing residential care.
- Fire safety in educational premises.
- Fire safety in small and medium places of assembly.
- Fire safety in large places of assembly.
- Fire safety in theatres and cinemas.
- Fire safety at outdoor events.
- Fire safety in hospital premises.
- Fire safety in the transport network.

Absolute and qualified duties

'ABSOLUTE'

Absolute means that there is no choice as to whether to take the actions specified by the duty. It does not matter if the company is a large multinational organisation with financial resources or a small charity organisation - action must be taken to comply with the duty.

For example "Article 5 (1) Where the premises are a workplace, the responsible person **must** ensure that any duty imposed by articles 8 to 22 or by regulations made under article 24 is complied with in respect of the premises".

Article 13 (3)(a) requires, where necessary, that the responsible person **must** "take measures for fire-fighting in the premises, adapted to the nature of the activities carried on there and the size of the undertaking and the premises concerned;". The responsible person has no choice and must take action to meet the duties expressed in the Article.

'PRACTICABLE'

Practicable means that the requirement must be carried out so far as it can be achieved within the current state of knowledge and intervention - even though implementation may be difficult, inconvenient and/or costly. For example, article 15 (2)(a) - "so far as is **practicable**, require any relevant persons who are exposed to serious and imminent danger to be informed of the nature of the hazard and of the steps taken or to be taken to protect them from it;" The requirement does not limit itself to a particular way of achieving this as it may vary as technology changes. It also means that in circumstances where the Article applies, organisations have to keep up with available technology in order to meet the requirement.

'REASONABLY PRACTICABLE'

A statutory duty which has to be carried out as far as is reasonably practicable is one where a balance of risk and costs is involved. An employer or 'responsible person' is entitled to balance costs of remedy against benefits in reduction of risk and if the benefit is minimal compared to the cost, he/she need not carry out the duty.

For example, article 8 (1)(a) requires that "the responsible person must take such general fire precautions as will ensure, so far as is **reasonably practicable**, the safety of any of his employees;". This is a very general duty and reflecting this general nature the responsible person is able to take action to comply based on a balance of risk and costs.

'REASONABLE'

Reasonable is a less onerous duty and though it is used in this context as a statutory duty it is only the same level of duty expected in common law related to negligence. An example of the statutory duty is Article 20 (3)(b) which states "the responsible person must take all reasonable steps to ensure that any person from an outside undertaking who is working in or on the premises receives sufficient information to enable that person to identify any person nominated by the responsible person in accordance with article 15 (1)(b) to implement evacuation procedures as far as they are concerned." This less onerous duty reflects the difficulty in achieving the requirement and the reliance on the co-operation of others to achieve it.

The powers of inspectors under the Regulatory Reform Order 2005

The enforcing authority specified by RRFSO is the Fire and Rescue Authority in the case of most premises; however the Order recognises that fire safety in some premises is a complex and technical area that has been enforced in earlier legislation by the Health and Safety Executive (HSE). The HSE enforce the RRFSO in nuclear installations, a ship under construction, conversion or repair and construction sites. The fire service maintained by the Secretary of State for Defence enforces the RRFSO for premises occupied solely for the purposes of armed forces of the Crown. The relevant local authority enforces the RRFSO in relation to premises requiring a safety certificate under the Safety of Sports Grounds Act or Safety of Places of Sport Act. A fire inspector authorised by the Secretary of State will enforce the RRO in Crown premises or the United Kingdom Atomic Energy Authority premises.

Inspectors are duly authorised officers of the enforcing authorities, they are provided with powers under the RRFSO to enable them to regulate compliance with the RRFSO. An inspector must, if so required, produce to the occupier of the premises evidence of his authority. Powers of inspectors authorised under the RRFSO are covered in Article 27 of the Order which sets out a general power to "do anything necessary for the purpose of carrying out this Order and any regulations made under it". In addition, the Order creates the following specific powers:

- To enter any premises to inspect it and anything in it, where this may be effected without the use of force.
- To make inquiries.
  - (i) To ascertain if the Order applies or has been complied with.
  - (ii) To identify the responsible person in relation to the premises.
- To require the production of any records (including plans).
  - (i) Which are required to be kept.
  - (ii) Which it is necessary to see for the purposes of an examination or inspection.
- To inspect and take copies of records.
- To require facilities and assistance.
- To take samples of any articles or substances for the purpose of ascertaining their fire resistance or flammability.
- To cause articles and substances that may cause danger to be dismantled or subjected to any process or test (but not so as to damage or destroy it unless this is, in the circumstances, necessary).

Where an inspector proposes to dismantle or test any article or substance, he must, if requested cause anything which is to be done by virtue of that power to be done in the presence of a person. Before starting to dismantle or test any article or substance an inspector must consult such persons as appear to him appropriate for the purpose of ascertaining what dangers, if any, there may be in dismantling or testing any article or substance.

Criminal liabilities

ENFORCEMENT NOTICES (IMPROVEMENT, PROHIBITION)

Part 3 of the RRFSO details the different types of enforcement notices that can be used, as follows:

Conditions for serving

Alterations notice
- Issued on premises at the discretion of an enforcing authority.
- Issued if premises constitute a serious risk, or may constitute a serious risk if changes were to be made to them or their use.
- If a notice is issued the person receiving must inform the enforcing authority of any proposed changes.

- Appeals can be made within 21 days of the serving of the notice to a magistrate's court, which suspends the notice till the hearing.

Enforcement notice

- This notice performs a similar function to an 'improvement notice' served under the Health and Safety at Work etc Act (HASAWA) 1974.
- Issued where there is a failure to comply with any of provisions of Order.
- Requires that person to take steps to remedy the failure within such period from the date of service of the notice (not being less than 28 days) as may be specified in the notice.
- The notice will specify the provisions that must be complied with and may include directions as to the measures which the enforcing authority considers are necessary to remedy the failure.
- The enforcing authority may withdraw the notice at any time before the end of the period specified in the notice.
- If an appeal against the notice is not pending, the enforcing authority may extend or further extend the period specified in the notice.
- If the notice would cause the premises to be altered the enforcing authority must consult any other relevant enforcing authority.
- Appeals can be made within 21 days of the serving of the notice to a magistrate's court, which suspends the notice till the hearing.

Prohibition notice

- Issued where the risk to 'relevant persons' is so serious that use of premises should be prohibited or restricted.
- The notice will specify the provisions that must be complied with and may include directions as to the measures which the enforcing authority considers are necessary to remedy the failure.
- The notice may direct that the use to which the prohibition notice relates is prohibited or restricted until the specified matters have been remedied.
- It may take effect immediately or at a time specified in the notice, depending on the opinion of the enforcing authority.
- The enforcing authority may withdraw the notice at any time before the end of the period specified in the notice.
- Appeals can be made within 21 days of the serving of the notice to a magistrate's court, the notice remains in force till the hearing.

In addition to the above, Article 37 in Part 5 (Miscellaneous) of the RRFSO deals with Fire-fighters' switches for equipment such as luminous discharge tubes. As part of this Article the fire authority can serve a notice on the responsible person regarding the location, colour and marking of the fire-fighters' switch.

Once the enforcing authority is of the opinion that conditions in or on the premises are such that any of the above notices can be served, they may choose to do so. However, they may need to consult with other enforcing authorities or the relevant local authority before issuing the notice, particularly if it required the premises to be altered *(see also - enforcement notice - on previous page)*. The notice can be served in person, by post, by leaving it at the responsible persons proper address, by electronic transmission or if there is no apparent responsible person, by affixing it to some conspicuous part of the premises.

Effects

Once a notice is issued they would normally have immediate effect unless stated otherwise. Being issued with a notice is not an offence; however non compliance with a notice becomes an offence.

Procedures, rights and effects of appeal

If a person has been issued with a notice they may appeal against the notice. Appeal is made by way of 'complaint for an order' at the Magistrate's Court and must be done within 21 days from the date that the notice was served. Once an appeal is commenced the 'alterations' and 'enforcement' notices are suspended. This however is not the case with a 'prohibition notice' which remains in force while the appeal is considered.

Role of Magistrates Court

The Magistrates Court is the level of court that is utilised to hear appeals against notices served under the RRFSO. However, cases may be referred to Crown Court, for example if a person appealing against a notice disagrees with the decision made by the magistrates court.

Penalties for failure to comply

If a notice is served on the 'responsible person' but they fail to comply with the notice this is seen as a separate offence to the original non-compliance that led to the notice being served. Article 32 sets out the following offences relating to failure to comply with notices:

- Fail to comply with any requirement imposed by an enforcement notice.
- Fail to comply with any prohibition or restriction imposed by a prohibition notice.

In each case failure to comply with a notice is considered to have serious potential and as such the penalty on summary conviction is the maximum that is allowed, presently £5,000. If the matter is taken on indictment it could lead to an unlimited fine, 2 years imprisonment or both.

PROSECUTION

Summary and indictable offences

Part 4 of the RRFSO details the offences that may occur related to the Order.

Offence	Summary	Indictment
Failure to comply with requirement or prohibition imposed by Articles 8 to 22 and 38 (fire safety duties) which places 'relevant persons' at risk of death or serious injury.	Statutory maximum (presently £5,000)	Unlimited fine, 2 years imprisonment or both
Failure to comply with regulations made under Article 24 which places 'relevant persons' at risk of death or serious injury.	Statutory maximum (presently £5,000)	Unlimited fine, 2 years imprisonment or both
Failure to comply with an alterations notice.	Statutory maximum (presently £5,000)	Unlimited fine, 2 years imprisonment or both
Failure to comply with an enforcement notice.	Statutory maximum (presently £5,000)	Unlimited fine, 2 years imprisonment or both
Luminous tube signs	Level 3 (presently £1,000)	Not applicable
Failure to comply with requirements imposed by Article 23 (general duties of employees at work) which places 'relevant persons' at risk of death or serious injury.	Statutory maximum (presently £5,000)	Unlimited fine
False entry in records etc.	Level 5 (presently £5,000)	Not applicable
Intentional obstruction of an inspector.	Level 5 (presently £5,000)	Not applicable
Fail to comply with any requirements imposed by an inspector.	Level 3 (presently £1,000)	Not applicable
Pretend with intent to deceive, to be an inspector.	Level 3 (presently £1,000)	Not applicable
Failure to comply with prohibition under Article 40 (duty not to charge employees).	Level 5 (presently £5,000)	Not applicable
Failure to comply with any prohibition or restriction imposed by a prohibition notice.	Statutory maximum (presently £5,000)	Unlimited fine, 2 years imprisonment or both

Part 4 also explains that the legal onus for proving that an offence was not committed is on the accused, and this is similar to that found in the HASAWA 1974. A new disputes procedure to deal with situations where a responsible person and an enforcing authority cannot agree on the measures which are necessary to remedy a failure to comply with the Order is also outlined within this part; this is via the Secretary of State's office.

Criminal courts

Cases are normally heard in a Magistrates Court where they are dealt with under summary proceedings. However, if the seriousness of the offence warrants it the case can be referred on to the Crown Court on indictment, where the proceedings allow for higher penalties to be awarded.

Penalties

Penalties awarded depend on the type of offence *(see also - Prosecution - above)*. The penalties are:

On summary conviction not exceeding 'statutory maximum'	-	presently £5,000
Level 3 on Standard Scale	-	presently £1,000
Level 5 on Standard Scale	-	presently £5,000
Fine on conviction on indictment:	-	unlimited fine
	-	imprisonment for up to 2 years
	-	unlimited fine and imprisonment for up to 2 years

Powers of authorised officers under the Fire and Rescue Act 2004

TO ENTER PREMISES FOR FIRE FIGHTING

Section 44 of the Fire and Rescue Act (FRA) 2004 provides authorised employees of a fire and rescue authority with the powers to deal with fires, road traffic accidents and other emergencies. It replaces section 30(1) of the Fire Services Act 1947 which was limited to dealing with extinguishing, or preventing the spread of, fires, and recognises the wider range of duties of fire-fighters, including the work which fire and rescue authorities do in responding to road traffic accidents.

It gives power to "do anything he reasonably believes to be necessary" in relation to extinguishing or preventing fires in an emergency and specifically to "enter premises or a place, by force if necessary, without the consent of the owner or occupier of the premises or place".

TO OBTAIN INFORMATION TO ASSIST IN FIRE FIGHTING AND EMERGENCIES

Section 45 of the FRA 2004 allows an authorised officer of a fire and rescue authority to enter premises at any reasonable time to obtain information that is needed for the discharge of the core functions of fire-fighting (clause 7), dealing with road traffic accidents (clause 8) and specified emergencies (clause 9). Such entry cannot be forcible and 24 hours notice must be given to the occupier of a private dwelling, unless authorised by a Justice of the Peace.

In exercising these powers an authorised officer can:

- Take with him any other persons and any equipment that he considers necessary.
- Require any person present on the premises to provide him with any facilities, information, documents or records, or other assistance, that he may reasonably request.

INVESTIGATE CAUSES AND PROGRESSION OF FIRES

Section 45 provides powers to authorised officers to enter premises at any reasonable time, if there has been a fire in the premises, for the purpose of investigating what caused the fire or why it progressed as it did.

Such entry cannot be forcible and 24 hours notice must be given to the occupier of a private dwelling, unless authorised by a Justice of the Peace.

In exercising these powers an authorised officer can:

- Take with him any other persons, and any equipment, that he considers necessary.
- Inspect and copy any documents or records on the premises or remove them from the premises.
- Carry out any inspections, measurements and tests in relation to the premises, or to an article or substance found on the premises, that he considers necessary.
- Take samples of an article or substance found on the premises, but not so as to destroy it or damage it unless it is necessary to do so for the purpose of the investigation.
- Dismantle an article found on the premises, but not so as to destroy it or damage it unless it is necessary to do so for the purpose of the investigation.
- Take possession of an article or substance found on the premises and retain it for as long as is necessary for any of the following purposes:

    (i) To examine it and do anything he has power to do.

    (ii) To ensure that it is not tampered with before his examination of it is completed.

    (iii) To ensure that it is available for use as evidence in proceedings for an offence relevant to the investigation.

- Require a person present on the premises to provide him with any facilities, information, documents or records, or other assistance, that he may reasonably request.

1.4 - The roles and functions of external agencies

FIRE AUTHORITY

Fire authorities have the responsibility for enforcing the RRFSO in the majority of premises. Under the Fire and Rescue Act (FRA) 2004 they are also responsible for:

- Promoting community fire safety, with the aim of preventing deaths and injuries in the home and reducing the impact of fire on the community as a whole.
- Planning and providing arrangements for fighting fires and protecting life and property from fires within its area and for receiving and responding to calls for help and for obtaining information to exercise its functions. The latter might include, for example, information about the nature and characteristics of buildings within the authority's area or availability of and access to water supplies.
- Rescuing persons from road traffic accidents and for dealing with the aftermath of such accidents.
- Responding to particular types of emergency, as defined by order, such as flooding and terrorist incidents.
- Equip and respond to events beyond its core functions provided for elsewhere in the Bill. A fire and rescue authority will be free to act where it believes there is a risk to life or the environment. This would allow, for example, specialist activities such as rope rescue.

- The fire and rescue authority may agree to the use of its equipment or personnel for any purpose it believes appropriate and wherever it so chooses. For example, a fire and rescue authority may agree to help pump out a pond as a service to its community.

HEALTH AND SAFETY COMMISSION (HSC)

On behalf of the government the HSC identifies the need for legal requirements, arrange drafts and consultation. They have an influence on law from the point of view that they decide what laws are appropriate and when they are to be introduced. As such they can control the quantity of law and the scope/extent of a given law. If organisations successfully lobby these bodies they may be able to gain modification or delay.

HEALTH AND SAFETY EXECUTIVE (HSE)

The HSE is appointed as an enforcing authority under the RRFSO for:

- Nuclear Installations.
- A ship, including Navy, which is in the course of construction, reconstruction or conversion or repair by persons who include persons other than the master and crew of the ship.
- Construction sites.

LOCAL AUTHORITIES

The local authority is appointed as an enforcing authority under the RRFSO for:

- A sports ground designated as requiring a safety certificate under section 1 of the Safety of Sports Grounds Act 1975[23] (safety certificates for large sports stadia).
- A regulated stand within the meaning of section 26(5) of the Fire Safety and Safety of Places of Sport Act 1987[24] (safety certificates for stands at sports grounds).

The local authority has a duty, under article 45 of the RRO, to consult the fire authority before passing plans to erect or extend a building, deposited with them in accordance with building regulations.

ENVIRONMENT AGENCY (EA) / SCOTTISH ENVIRONMENTAL PROTECTION AGENCY (SEPA)

The Environment Agency is a Non-Departmental Public Body concerned with protecting and improving the land, air and water environment of England and Wales. The Scottish EA has a similar role in Scotland.

As the RRO imposes a duty on the 'responsible person' to "mitigate the effects of a fire on anyone on the premises, and on anyone in the vicinity of the premises that a fire on the premises may effect", any fire that causes pollution will be of interest to the Environment Agency. Fire Authorities and the Environment Agency usually work in conjunction with each other, with environmental damage limitation equipment being made available to the Fire Authority by the Environmental Agency.

INSURANCE COMPANIES

Insurance companies have become increasingly aware that they may have undervalued the risks related to some companies they insure. This has caused them to review the factors that lead to claims. In conjunction with this they will influence organisations to improve their fire safety management, fire prevention and fire protective measures so as to reduce the level of risk. This new approach includes an assessment of the impact any changes to the local Fire and Rescue Service's planned responses to any fire incident, or other emergency, involving the insured property.

1.5 - Sources of information on fire safety risks and control

Information on fire safety measures, fire risk assessment and control measures that are type specific to the premises type can be found in the guidance books issued by the Office of the Deputy Prime Minister (ODPM). There are 11 of these guides which are relevant to:

- Fire safety in offices and shops.
- Fire safety in factories and warehouses.
- Fire safety in premises providing sleeping accommodation.
- Fire safety in premises providing residential care.
- Fire safety in educational premises.
- Fire safety in small and medium places of assembly.
- Fire safety in large places of assembly.
- Fire safety in theatres and cinemas.
- Fire safety at outdoor events.
- Fire safety in hospital premises.
- Fire safety in the transport network.

However there are many sources of information that can be of assistance to the 'responsible person'. There is now a lot of information freely available on the internet and a selection of relevant websites would be:

- Http://www.arsonpreventionbureau.org.uk/
- Http://www.bre.co.uk

- Http://www.firekills.gov.uk/leaflets/index.htm
- Http://www.firesafetytoolbox.org.uk/ncfsc/default.htm
- Http://www.hse.gov.uk
- Http://www.ife.org.uk
- Http://www.means-of-escape.com
- Http://www.odpm.gov.uk
- Http://www.thefpa.co.uk

1.6 - Fire safety management

A framework for fire safety management

The RRFSO imposes a legal duty on the responsible person to put into place safety arrangements. The arrangements must take into account the size and nature of the activities. The scope of the arrangements must include planning, organisation, control, monitoring and review. In practice the scope of arrangements required reflects the common components of good management. Health and safety legislation has required this approach to be taken for matters of general health and safety for some time, the RRFSO puts particular emphasis on fire safety arrangements.

SETTING POLICY

Though the RRFSO does not require an organisation to set out its policy on fire safety specifically, a typical framework to implement fire safety would start with the organisation's fire safety policy. As with general health and safety it is important that the organisation and its senior management make a commitment to prevention. This may be as part of an overall health and safety policy, integrating fire safety within the policy statement or as a separate statement of commitment. It is good practice to give a senior manager the overall responsibility for fire safety management. However it is worth remembering that the 'responsible person' cannot absolve their responsibility.

ORGANISING

The RRO establishes responsibility to the "responsible person". In the case of a workplace this is the employer if the workplace is to any extent under the employer's control. In the case of other premises it is either the person in control of the premises, such as an occupier, in connection with a trade, business or other undertaking, or the owner where the person in control does not have control in connection with a trade, business or undertaking - for example, for common parts of an office block rented out to various organisations.

The responsible person must appoint a sufficient number of competent persons to assist with undertaking the preventive and protective measures required by the RRFSO. In addition, where more than one is appointed arrangements must be made for ensuring adequate co-operation between them.

The responsible person must implement appropriate arrangements for the effective planning, organisation, control, monitoring and review of the preventive and protective measures. In addition, the responsible person must record these arrangements:

- If he employs five or more employees.
- If a licence under an enactment is in force in relation to the premises.
- If an alterations notice is in force in relation to the premises.

The arrangements should be flexible enough to allow for change and may cover such areas as:

- Responsibility for fire safety at board level.
- Responsibility for each premises (usually the site manager or person who has overall control).
- Arrangements for appointing people to carry out specific roles in the event of a fire.

The responsible person must obligate and enable line management to implement the fire safety arrangements put in place. Care should be taken to develop their competence in fire safety management and co-ordinate their effort. Roles and responsibilities should be clearly defined in writing and relationships expressed on an organisational chart. The organisation must ensure adequate arrangements are in place to ensure consultation with employees on matters of fire safety.

PLANNING AND IMPLEMENTING

A successful organisation will adopt a planned and systematic approach to policy implementation. Their aim must be to minimise the risk of fire starting and the effect of fire on both people and the business. They should use fire risk assessment methods to enable focused planning and the outcomes of the risk assessments should be used to prioritise and implement the preventive and protective measures that are needed. Planning and implementation should include:

- Identification of fire hazards and the assessment of risks that arise from them.
- A plan of action should be compiled as a result of the fire risk assessment and the measures that are needed should be prioritised relevant to risk.
- Fire prevention and protection measures and procedures in place.
- Procedures for fighting fire.
- Identity of people with specific duties in the event of a fire.
- Details of dangerous substances.

- Written instructions to staff.
- Co-operation and co-ordination between occupiers.
- Fire safety training.
- Fire drills.
- The system for monitoring that requirements are being met.
- The system for fire safety records.
- A system for providing information to employees, contractors, visitors, other persons as necessary and the emergency services needs to be implemented.

MONITORING AND REVIEWING

It is essential that the intended implementation of arrangements be put in place to monitor and to ensure compliance with intention and to ensure their effectivness. This should not solely rely on reactive monitoring through measurement of the number of reported fire incidents, but should include proactive means of monitoring. Successful organisations use a mixture of proactive methods, including inspections and meetings to confirm actions and progress.

The arrangements, and therefore the fire management system, that are implemented should allow for and react to changes in the premises, people, materials and processes, or in fact to any change (or proposed change) that may necessitate an amendment to the specific arrangements or fire management system in place. In order to do this planned reviews should be organised to follow fire incidents, proposed changes affecting the premises and after a period of time if not done for another reason.

AUDITING

It is imperative that a process of auditing fire safety arrangements and the system of fire management be periodically undertaken. This will determine the degree to which the organisation is in compliance with its own fire safety standards and provide an opportunity to identify if these standards are effective. An audit must examine the standards and compliance with them within the management system for fire safety. It is more than a physical conditions inspection. Although inspections are a useful way of monitoring fire safety, they are not a substitute for a comprehensive audit conducted by a competent auditor. The frequency of audit should depend on the level of risk relating to the premises organisation.

Fire safety management and the protection of people from fire

Fire safety should be approached in a similar manner to health and safety. Management should look to create the culture so that the 'fire safe person' is created. Fire safety management operates at all levels within an organisation and each individual should know and understand what role they play in the fire safety management within the premises. A good management system will ensure that as any fire safety issues arise the individuals have the knowledge, skills and flexibility within the system to manage the issue locally without taking risks relating to the fire safety measures that are needed.

Duties under the Regulatory Reform (Fire Safety) Order 2005

PLANNING

Article 11 gives a direct requirement for the responsible person to put into effect the effective planning of the preventive and protective measures.

ORGANISATION

Article 11 gives a direct requirement for the responsible person to put into effect the organisation of the preventive and protective measures.

CONTROL

Article 11 gives a direct requirement for the responsible person to put into effect the effective control of the preventive and protective measures.

MONITORING AND REVIEW OF THE PREVENTIVE AND PROTECTIVE MEASURES

Article 11 gives a direct requirement for the responsible person to put into effect a system for monitoring and review of the preventive and protective measures.

The structure of the duties and burden of proof requirements of the RRFSO make post-fire incident investigation and enforcement action a real possibility. The responsible person can only defend any charges made by demonstrating that the actions that they took and the manner in which they managed the risks to people were in accordance with the duties under the RRFSO. They will have to show that they took all due dilegence to avoid the commission of such an offence.

1.7 - Post fire management

Requirements for recording

STATUTORY AND NON-STATUTORY REQUIREMENTS

If an organisation is unfortunate enough to suffer a fire there will need to be certain actions taken dependant upon the scale of the fire and if any injury or death has occurred. Responsible persons need to ensure that fires are reported for the following reasons:

Statutory

- The report will help the 'responsible person' decide if the fire is reportable under the Reporting of Injuries Diseases and Dangerous Occurrences Regulations (RIDDOR) 1995 because somone was injured or killed or it constitutes a dangerous occurrence.
- It is an implied requirement of the RRFSO Article 11 to have arrangements to monitor preventive and protective measures.
- To provide evidence in any legal action that may be taken.
- The 'responsible person' must comply with the requests of the fire service investigating officer and supply such assistance and information as required.

Non statutory

- The report would enable an investigation.
- The investigation in turn, should help to identify flaws with existing controls and therefore assist in the implementation of improved controls.
- Analysis of reports may identify trends or patterns.
- Gathering statistical data will help the responsible person and, if the Fire and Rescue Service request the data may assist in development of national statistics to identify trends and comparisons.

RECORDING INCIDENTS, INJURIES AND DANGEROUS OCCURRENCES

Accident book

If any injury has occurred due to a fire, for example, harm due to smoke inhalation, it should be reported and recorded in the accident book like any other injury at work.

Fire log book

A log book should be kept to record the outcomes from fire prevention arrangements in place. It should be kept with or contain records of all testing, maintenance, training, drills and audit records.

General incident or occurrence book

General incident or occurrence book should be kept on site and available to record the various events that may occur relating to fires. This is particualrly important if an organisation suffers from fire alarm operations, near misses or fires on a frequent basis. In this way any trends or patterns may emerge, which may assist in improvements to the fire management system.

PROCESS AND PROCEDURES FOR REPORTING

Fire related fatalities

Using their powers under the FRA 2004 the fire and rescue authority would undertake an investigation if anyone is killed as a result of a fire. In addition to this there is a responsibility on the employer to comply with RIDDOR 1995 requirements. This will involve notification and reporting of the death to the enforcing authority realted to RIDDOR 1995, either the Health and Safety Executive (HSE) or Local Authority, as applicable.

It should be remembered that the duty under RIDDOR is on the worker's employer to notify and report under RIDDOR. If the person is self employed or a member of the public the duty to notify or report lies with the organisation in control of the undertaking from which the death resulted.

Major injuries

Just as with deaths resulting from a fire, the fire and rescue authority would undertake an investigation if anyone is seriously injured. The duties to report a major injury resulting from a fire are the same as if a death had occurred.

Dangerous occurrences

Dangerous occurrences relating to fire that are reportable under RIDDOR 1995 are:

- Explosion, collapse or bursting of any closed vessel or associated pipework.
- Electrical short circuit or overload causing fire or explosion.
- A road tanker carrying a dangerous substance overturns, suffers serious damage, catches fire or the substance is released.
- A dangerous substance being conveyed by road is involved in a fire or released.

- Explosion or fire causing suspension of normal work for over 24 hours.
- Sudden, uncontrolled release in a building of: 100 kg or more of flammable liquid; 10 kg of flammable liquid above its boiling point; 10 kg or more of flammable gas; or of 500 kg of these substances if the release is in the open air.

There is a legal requirement for certain fire incidents to be reported, whether people are hurt or not.

In each type of event above, death, major injury or dangerous occurrence, must be notified to the enforcing authority, either the HSE or Local Authority, by the quickest practicable means and a written report in the form of an F2508 provided within 10 days of the event.

NEED TO REVIEW AND REVISE FIRE RISK ASSESSMENTS

It is essential that after a fire the arrangements to manage the prevention of fire be reviewed to determine their effectiveness. This will include any risk assessments conducted. It is important to relate the fire and the success or failure of the preventative and protective measures to the original risk assesssment and determine if the controls identified as needed were put in place and whether they were adequate. The lessons learned for this risk assessment should be used to review other similar risk assessments or risk assessments that relied on similar controls.

See also - Fire Risk Assessment - Element 6.

Fire investigation

THE PURPOSE OF INVESTIGATING FIRES

The purpose of fire investigation will vary depending on the standpoint of the investigator. For example the responsible person should investigate to determine the cause and enable preventive measures to be put in place for the future. In addition the responsible person may be aware of the prospect of civil or criminal litigation and may also want to gather evidence in anticipation of this. A Fire and Rescue Authority fire officer may carry out an investigation to determine the cause, but may also want to gather evidence in order to lay a criminal charge against the responsible person. An insurance company investigator may investigate to determine if they have liability for any claim arising from the fire.

BASIC FIRE RELATED INVESTIGATION PROCEDURES

Non fatal fires - accidental or arson, false alarms

It is essential to take full advantage of all fire events and it is particullarly important to learn from those events that have not yet caused harm to people, as they may indicate a pattern of failure in arrrangements and will provide an opportunity to make improvements before a death ocurrs. These events provide a valid test of preventative and protective arrangements along with an opportunity to review the approach taken. If accidental false alarms persist it can undermine the importance of the fire safety arrangements put in place, so it is important to investigate and remedy the problem promptly and communicate the outcome to those affected in order to restore confidence in the arrangements.

Once a fire is extinguished the Fire and Rescue Authority fire officer in charge of the incident would initiate a fire investigation to ascertain the cause of a fire. On occasions a specialist fire safety officer will take over, usually when more time and expertise is required. The resulting information is forwarded to the appropriate government department which compiles national statistics. When these statistics are analysed they can identify areas where fire prevention measures can be introduced to reduce fire losses.

THE PROCEDURAL DIFFERENCES AND DEFINITIONS

Accidental fire investigations

If arson is not suspected the fire would be considered to be accidental. This does not mean that it is without cause or that someone cannot be held responsible for it. The enforcing authority would use powers under the Fire and Rescue Act and/or the RRFSO to determine the causes of the fire and determine if there were breaches sufficient for enforcement action. The police do not need to be involved in these investigations as the enforcing authority has the powers to proceed to prosecution. Fire investigation is divided into various stages:

- Interviewing eye witnesses.
- Locating the seat of the fire.
- Excavating the seat.
- Evaluation of evidence.
- Review all findings.
- Report.

The above sequence would be the preferred method, but occasionally the investigation may have to be conducted out of sequence.

Arson set fire investigations

The fire service also categorise non-accidental fires as either 'malicious', 'deliberate' or 'doubtful'. The term 'doubtful' can be misinterpreted; 'doubtful' does not mean that the fire service does not know the cause, but that the fire is suspicious. For the cause of a fire to be recorded as 'doubtful', deliberate ignition has only to be suspected, not proven.

Arson or malicious firing is investigated by the fire and rescue authority to assist the police in apprehending the perpetrators. It is essential that a properly conducted formal fire investigation is undertaken to identify the cause. A specialist officer will be involved and they must work in partnership with the police, which may involve police forensic scientists.

The police are bound by the Criminal Damage Act 1971, which means in order to record an offence of arson they have to prove that persons behaved 'recklessly' or 'intended to damage property'. In English law malicious firing is only considered arson if a person's life is put in danger as the result of the fire. In Scottish law they do not have malicious firing - it is considered a degree of arson.

INVESTIGATION PREPARATION

The investigation may start before the fire is extinguished which may have an effect on the fire-fighting methods used.

Preserving the fire scene

One factor in all cases where a fire investigation is to take place is the importance of not disturbing the scene of fire any more than is necessary during fire-fighting. If the fire results in a death, or arson is suspected, the fire fighting may be stopped or amended so as to allow an initial investigation to take place before the evidence is ruined.

LIAISON WITH THE POLICE, FIRE OFFICER, HSE, PUBLIC UTILITIES AND INSURANCE INVESTIGATORS

The responsible person should carry out their own investigation, to determine the effectiveness of their fire arrangements/management system. The responsible person would need to liaise with a variety of external organisations, depending on the nature of the fire incident and the premises in which it takes place. Liaison will need to take place with other persons/bodies for a variety of reasons. Examples of this could be:

- **Police** - The police are responsible for the criminal investigations on arson or suspicious fires. It will be necessary to liaise and assist them (and the Fire and Rescue Service) following any fire caused by arson. It may also be necessary to talk to the police as part of undertaking risk assessments, regarding security issues.
- **Fire officer** - As above the fire officer has the power to investigate fires and to take samples for investigation purposes. There is a duty to assist the officers in there investigations. It may also be necessary to speak to the fire officer if you have dangerous substances on site to discuss emergency actions, or to discuss fire appliance attendances and other issues which may impact on a fire risk assessment. For large premises with quantities of building slabs with flammable infills it may also be necessary to discuss this with the Fire and Rescue Service so they can make an assessment of the risks to their staff and develop their operational strategy as a result.
- **HSE** - It may be necessary to discuss any process fire risks with the HSE.
- **Public utilities** - It may be necessary to discuss issues re drainage, water pollution, and fire water run offs with organisations such as the water boards or environmental agencies. This may be the case for example if you have dangerous substances on site, due to the potential for pollution in the event of a fire and your duty to mitigate the effects of a fire.
- **Insurance investigators** - If an organisation is unfortunate and suffers a fire it will need to liaise with its insurers and loss adjuster. It may also be necessary to speak to the insurer whilst undertaking fire risk assessments to ascertain the implications of wording in insurance documentation and the importance of following the guidelines set. As an example: if a policy asks for combustibles to be stored 8m away from the premises, but they are not, and a fire in the rubbish skip spreads to the premises itself the insurers may not settle a claim that is made for loss.

IDENTIFYING THE REASON FOR, OR UNDERLYING CAUSE OF, THE FIRE

There are many reasons why fires are investigated including detection of crime, verification of insurance claims, prevention of future fires, identification of defective components and dangerous processes. The level and type of investigation may be dictated by the type of incident and its effects.

Fires will be caused either accidentally or deliberately (arson) and they may cause injuries (non-fatal fires), or deaths (fatal fires). Specialist Fire and Rescue Service fire safety officers will normally investigate fires that have resulted in the death or injury of a person and the subsequent report is submitted to the coroner's office (in the case of a death), for the inquest.

False alarms may also be investigated by the Fire and Rescue Authority in order to determine the cause of the alarm. However, if an excess number of false alarms is being created by a premises, the investigation may be done to decide on potential action, which may include a reduction of the level of response by the Fire and Rescue Authority. As will be discussed in Element 3, a three level response to fire alarms is being considered, with the level of response given based on the number of false alarms the fire alarm system generates.

Trigger points will be set and once a premises number of false alarms goes above this point the Fire and Rescue Service will investigate. If it is found for example that the calls that are generated are preventable, e.g. toaster or dust from work process sets off alarm, the Fire and Rescue Service will work with the responsible person to manage down the level of calls created. If the responsible person does not comply with this request, or the fire management within the premises is such that the number of calls does not reduce, then the premises will be lowered to the next level of response.

Remedial actions

Following the investigation of a fire, areas where fire prevention measures can be introduced to reduce the risk of fires starting may be highlighted. If the investigation is being done from a false alarm, remedial actions may need to be taken by the Responsible Person to prevent the fire and rescue authority lowering the standard of cover that is given to the premises.

Any fire or near miss must be investigated as part of the fire risk assessment review process. Any incident investigation will in all probability highlight areas of weakness in the fire management system that need to be addressed. In the context of a legal system where we have a duty to prevent fires, it is paramount that we review any incident that has occurred and do everything that is 'reasonably practicable' to prevent such an event happening again.

SITE AND DAMAGED AREA CLEAN UP CONSIDERATIONS

During the clean up of both the site and damaged area it is paramount that the health and safety of those involved is ensured. It is essential to realise that after a fire the structure may be unstable, services exposed and substances released. The site will have similar features to a building left unoccupied for a long time. If such a building was going to be demolished a pre-demolition survey would be conducted. A similar approach should be taken with a fire damaged area in that a post fire survey should be conducted to evaluate the conditions and risks arising form the fire affected site. Specialist companies can be utilised to clean up equipment following a fire. This is especially important when considering the damage done to any high tech equipment - computers or any other sensitive equipment that may be damaged by the effects of the fire.

Principles of fire and explosion

Overall Aims

On completion of this Element, candidates will understand:

- combustion processes.
- ignition of solids, liquids and gases.
- fire growth and spread.
- explosion and explosive combustion.

Content

Specific Intended Learning Outcomes

The intended learning outcome of this Element are that candidates will be able to:

2.1   outline the principles behind ignition and spread of fire and explosion in work premises

2.1 - Combustion process

The concept of the fire triangle

A simple approach depicts fire as having three essential components - fuel, oxygen and heat. When the three separate parts are brought together a fire occurs. This is often depicted by the 'fire triangle'. This traditional approach is useful when considering the 'ingredients' needed to make a fire.

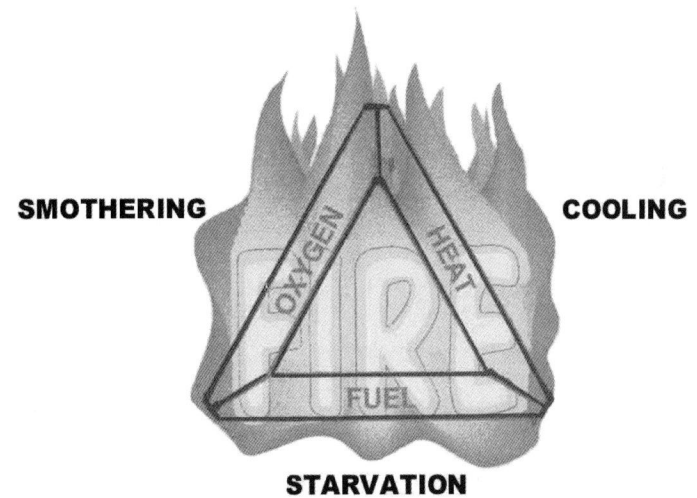

Figure 2-1: The fire triangle.

Source: CorelDRAW! 5.0 Clipart.

Combustion

THE CHEMISTRY OF COMBUSTION

Combustion is defined as being a chemical reaction during which heat energy and light energy are emitted. If the above three components form together in the right proportions, then the chemical reaction of combustion takes place.

THE CONDITIONS FOR THE MAINTENANCE OF COMBUSTION

FUEL   The fuel dictates all aspects of fire. The type of fuel, the format the fuel is in, or how much is present in the air will control its susceptibility to fire.

HEAT   The level of heat required to cause ignition is dictated by the format (solid, liquid, vapour or gas) and type of fuel.

OXYGEN   This is provided from the air around us. Other than in specialised industrial applications it is always present. Consideration should be given to how the available oxygen might inadvertently be increased (oxygen enrichment) via oxidising agents, or oxygen cylinders.

For the combustion process to be maintained all three of the component parts above must remain present.

If one or more of these parts of the fire is removed, the fire will be extinguished. This can be done by:

1.   COOLING the fire to remove the heat, e.g. water or foam extinguisher.
2.   STARVING the fire of fuel, e.g. isolation of gas supply.
3.   SMOTHERING the fire by limiting its oxygen supply, e.g. foam extinguisher.
4.   CHEMICAL INTERFERENCE of the flame reactions, e.g. method by which some extinguishing media works.

COMBUSTION AS A COMPLEX DYNAMIC PROCESS

Whilst the combustion process is often defined as a chemical reaction (or series of reactions) where heat and light energy are evolved, this is a very simplistic approach. A more accurate approach is to describe the chemical reactions in the combustion process as 'happening very quickly, at very high temperatures, and in very small volumes'. The tiny reactions that take place are exothermic and give out sufficient excess heat to initiate the next chemical reaction. Combustion is therefore a chain reaction, or indeed a series of branching chain reactions. As an example of the complex nature of combustion consider the burning of hydrogen.

This reaction is denoted chemically as: $2 H_2 + O_2 = 2H_2O$

This is a simple depiction of the reaction process. The equation implies that two molecules of hydrogen combine with one molecule of oxygen to form two molecules of water when they collide. Collisions like this must happen in controlled chemical processes but they do not explain the extremely rapid reactions that occur in a hydrogen/oxygen explosion.

All flame reactions depend on the activities of tiny fragments of molecules that arise from the reaction. These tiny fragments are very reactive, very unstable, and can only exist for a fraction of a second. However, they are very mobile and capable of rapid reproduction. These fragments are known as 'free radicals'.

If we consider the equation again we can see that the component parts can be broken up into hydrogen atoms (H), oxygen atoms (O), and hydroxyl radicals (OH). Inside a flame, all three components are very hot, in high concentrations and completely free to react.

Description of the stages of combustion

Fire tends to grow in stages. The graph shows that fires start with a slow induction period, which is where the conditions for fire are not quite right for ignition. For example, heat may be present and exposed to a source of fuel, but not sufficiently long enough to raise the temperature of the fuel to a high enough level for ignition to take place. Once ignition is reached combustion grows very quickly and is highly dependent on the level of oxygen in the area of the fire. It then reaches a steady state where heat available and the fuel/oxygen used is balanced. Once the fuel is consumed, the fire decays.

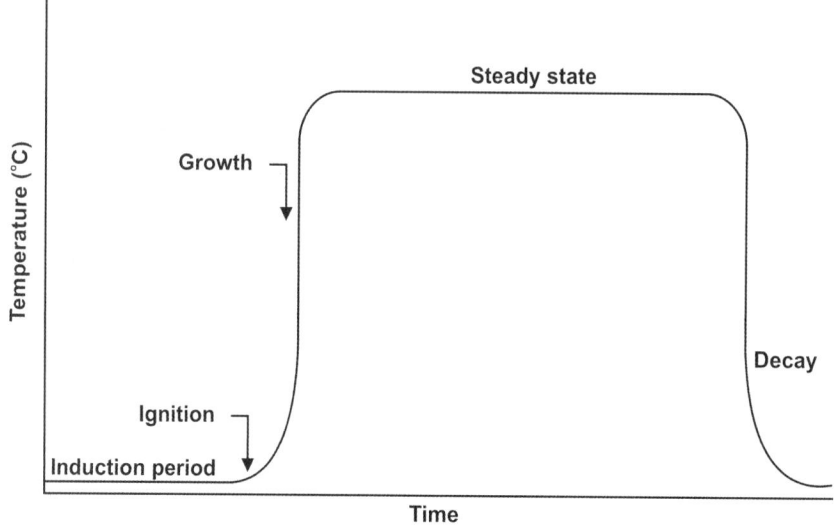

Figure 2.2: Stages of combustion.   *Source: ACT.*

INDUCTION

This represents the early stages of a fire, when the combustion process is starting. This may often be a slow process such as in a smouldering fire, where the chemical decomposition of a material and the heat build up to an ignition point may take some considerable time.

GROWTH

Once true combustion and ignition have taken place, this phase will often grow very quickly and is only limited by the availability of oxygen. The growth rate will then reach a plateau and will remain at that level until the fuel is consumed.

DECAY

This is the final stages of a fire. As the fuel runs out the combustion process will quickly slow down, and eventually come to a stop.

Outline of basic chemical reactivity

ENDOTHERMIC

Chemical reactions that must absorb energy in order to proceed are called endothermic reactions. Endothermic reactions cannot occur spontaneously, work must be done in order to get these reactions to occur. When endothermic reactions absorb energy, a temperature drop is measured during the reaction.

EXOTHERMIC

Many chemical reactions release energy in the form of heat, light, or sound. These are exothermic reactions. Exothermic reactions may occur spontaneously. Some exothermic reactions produce heat very quickly resulting in an explosion. The most common examples of exothermic reactions would be with fires involving organic peroxides. These products are dangerous in a fire as they readily give off quanities of oxygen. The chemical process taking place for this to happen gives out quantities of heat. If these products are affected by a fire the entire process is accelerated.

2.2 - Ignition

Identification of ignition sources

The majority of fires need an ignition source to enable them to start. It is imperative therefore that an assessment should be made in the workplace to identify all ignition sources likely to be present. Once these ignition sources have been located, then the surrounding areas can be checked for combustible materials or gases that are normally, or are likely, to be present.

Having now identified all problem areas a study can be made to assess the viability of separating the ignition source and combustible material from each other. If this is not possible then some form of control measure should be taken to minimise the risk of fire. HOLISTIC APPROACH - HAVE I GOT FUELS. CIGARETTE. HOT ENOUGH TO IGNITE MATERIALS IN CONTACT WITH. STEAM PIPE

Figure 2-3: Mobile ignition source. *Source: FST.*

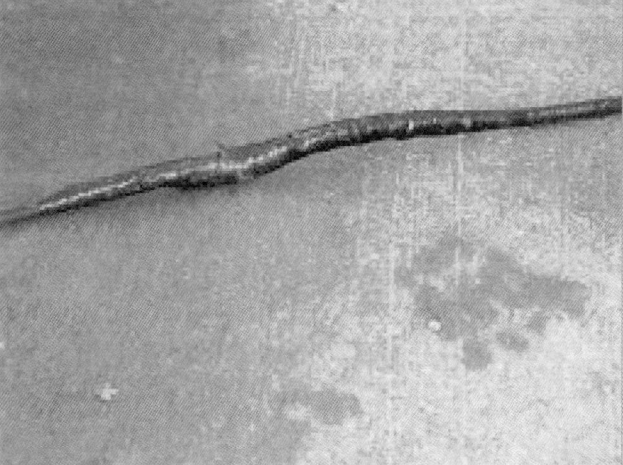

Figure 2-4: Defective electrical equipment. *Source: ACT.*

Ignition sources can be found in many forms. Some of the more common examples are listed below:

- Smokers' materials.
- Sparks from welding equipment.
- Sparks from electrical motors.
- Sparks from grinding equipment.
- Oxy-acetylene welding equipment.
- Fixed or portable heaters.
- Cooking equipment.
- Bitumen boilers.

- Electrical faults and overloaded electrical circuits.
- Steam pipes.
- Overheating equipment.
- Static electricity.
- Non-intrinsically safe equipment used in a flammable atmosphere.
- Radiated heat from a legitimate source such as a light bulb.
- Hot surfaces such as soldering irons or hot glue guns.
- People with the intent to cause harm or deliberate acts of ignition.

It is also important that good housekeeping is maintained. If any of the above ignition sources are present and the housekeeping is poor, then the likelihood of a fire breaking out is dramatically increased.

There are various types of ignition sources. The measures to control them are diverse.

The ignition of solid materials

The ease with which a solid material will ignite will depend upon the type of material and the way the material is presented. In simplistic terms it can be said 'the smaller the particles of material, the easier it is to ignite'. As a simple analogy, an A4 pack of paper will not ignite readily, as it is a solid mass with no air in the middle. However, separate the paper into individual sheets and it will ignite more easily.

It is important to note that the solid material does not actually burn. When a combustible solid is involved in a fire, it first chemically decomposes and produces carbon products in the form of a vapour. It is these that ignite when mixed with the oxygen from the air. This can be demonstrated by holding a match to a piece of paper. It will be seen that the paper does not burst into flames, but chars and goes black, vapours are given off and it is these vapours that burn.

The ignition of liquids and gases

Figure 2-5: Poor housekeeping. *Source: FST.*

A flammable liquid gives off vapours and it is these vapours that ignite. The temperature and rate at which vapours are given off will vary from one flammable liquid to another. Flammable liquids that more readily release vapours are said to be more volatile than others. Some of the volatile flammable liquids, for example, petroleum, readily give off vapours at room temperature. Others, such as diesel oil need to be heated to above room temperature before they give off sufficient vapours to be a significant risk. The influence of the temperature of a flammable liquid on ease of ignition is important and when measured is called the *'flashpoint'* of a flammable liquid. Once sufficient vapours have been given off, so that the vapour / air mixture is within the *'flammable limits'*, only a very small ignition source is required to cause ignition and for combustion of the vapours to take place over the surface of the liquid. Once combusting, the flammable liquid will generate more vapours because of the increased heat. The combustion process will continue until action is taken to extinguish the fire or the fuel runs out. Flammable liquids will spread fires very quickly, due to their tendency to rapidly flow across a surface and the rapid increase in heat energy generated by the combustion process.

Flammable gases behave differently to liquids as they are already in the gaseous state. These small gaseous particles readily mix with the available oxygen and are easily ignited by an ignition source. If the flammable gas is present, within its flammable limits and in the presence of an ignition source, a fire/explosion will occur throughout the airspace volume that the flammable gas occupies.

THE CONDITIONS FOR SUCH IGNITION TO OCCUR

Ignition occurs when a heat source e.g. a spark, contains sufficient heat energy to cause combustion of one or more molecules of a flammable vapour or substance.

THE METHODS OF AVOIDING SUCH IGNITION

The simple principle to avoid ignition is to separate the heat and fuel sources, or to manage the situation in some other way. Examples of this have been given earlier. Further examples are given below.

Hazard	Control measure
Smoking materials	Good housekeeping Management control e.g. defined areas for smokers
Sparks from welding gun	Hot work permits Management control, e.g. permit to work
Sparks from machinery	Good maintenance programmes Good housekeeping Correct choice of machinery
Static electricity	Safe system of work. Earthing straps/bonding Use of anti-static materials
Electrical faults	Competent staff. Good maintenance checks. Portable appliance testing.
Overheating machinery	Good maintenance programmes. Correct siting for airflows etc.
Radiated heat from legitimate source	Space separation. Shielding / fire resistance.
Emission of flammable vapours	Seal container.
Flammable vapours present in workplace	As above including electrical equipment to be approved for use in flammable atmospheres.

Figure 2-6: Source of ignition and control measures.   *Source: FST.*

Definitions

FLASH POINT

'Flash Point' is defined as the lowest temperature at which, in a specific test apparatus, sufficient vapour is produced from a liquid sample for momentary or flash ignition to occur. It must not be confused with ignition temperature which can be considerable lower.

AUTO IGNITION TEMPERATURE

'Auto Ignition Temperature' is the lowest temperature at which a substance will ignite spontaneously, and will burn without a flame or other ignition source. It is sometimes called 'Spontaneous-Ignition Temperature'.

VAPOUR DENSITY

This term is sometimes used in the 'fire world' and can cause confusion, as it is often mistaken for the 'density of a vapour' (relative density). Vapour density is the density of a gas or vapour compared to the density of hydrogen. This is worked out by using the molecular weights of the atoms concerned. Therefore the vapour density for oxygen, as an example, is

$$\frac{32}{2} = 16$$

(Molecular weight oxygen = 32, hydrogen = 2).

This figure of vapour density is in a way therefore of little value, as it is a theoretical comparison to hydrogen.

For fire safety and fire protection reasons we are concerned with the comparison of a material and its vapours in relation to air (in the majority of cases). The vapour density of air at standard temperature and pressure is taken as 22.4. We can see therefore that the **relative density** for oxygen when compared to air is:

$$\frac{32}{22.4} = 1.43\,g/l\,a\,stp\,(i.e.\,approximately\,1\tfrac{1}{2}\,times\,heavier\,than\,air).$$

As can be seen the difference is considerable and it must be ensured that the relative density of a vapour is used when considering its hazards and the necessary safety measures.

VAPOUR PRESSURE

'Vapour Pressure' is the pressure of a vapour given off by (evaporated from) a liquid or solid and is caused by atoms or molecules continuously escaping from its surface. In an enclosed space, a maximum value is reached when the number of particles leaving the surface is in equilibrium with those returning to it - this is known as the **saturated vapour pressure** or equilibrium vapour pressure.

FLAMMABLE LIMITS

A flammable gas or vapour will only burn in air if the mixture lies between certain limits. This is known as its flammable range. These limits are normally given as a percentage of the substance relative to air and are called:

- Upper Flammable Limit - the highest mixture of fuel and air that will just support a flame.
- Lower Flammable Limit - the lowest mixture of fuel and air that will just support a flame.

Product	Lower flammability limit	Upper flammability limit
Butane	1.9%	8.5%
Acetylene	2.5%	81%

Figure 2-7: Flammable limits. *Source: Institute of Fire Engineers.*

Remember, it is the vapour of a substance that burns. A solid or liquid must be heated to a temperature where the vapour given off can ignite before combustion takes place.

Description of the classification of fires

A basic understanding of the classes of fire needs to be taught because many fire extinguishers state the classes of fire which may be attacked using them. There are 5 classes into which fires can fall:

CLASS A		Fire involving solids (wood, paper, plastics, etc., usually of an organic nature).
CLASS B		Fires involving liquids or liquefiable solids (petrol, oil, paint, fat, wax, etc.).
CLASS C		Fires involving gases (liquefied petroleum gas, natural gas, acetylene, etc.).
CLASS D		Fires involving metals (sodium, magnesium and many metal powders, etc.).
ELECTRICAL HAZARDS		Although not a true class of fire, we should also consider fires in electrical equipment. This is a classification used with extinguishers to identify their suitability for use on electrical equipment.
CLASS F		Fire involving cooking fats and oils.

Figure 2-8: Classes of fire. *Source: Rivington Designs.*

2.3 - Fire growth

The factors that influence fire growth rate

OXYGEN CONTENT

One of the main factors that affect the fire growth rate will be the available amount of oxygen. If a fire occurs in a well ventilated room, it can grow at a very rapid speed.

Oxygen enrichment of the atmosphere, even by a few percent, considerably increases the risk of fire and fire growth rate. Sparks which would normally be regarded as harmless can cause fires and materials which do not burn in air, including fireproofing materials, to burn vigorously (or even spontaneously) in oxygen enriched air.

The level of oxygen available to a fire may be maintained or increased by ventilation equipment, for example, air conditioning systems that supply air to an office which could influence greatly the fire growth rate.

Oil and grease are particularly hazardous in the presence of oxygen as they can ignite spontaneously and burn with explosive violence. They should never be used to lubricate oxygen or enriched air equipment (special lubricants which are inert when used with oxygen should always be used).

Many accidents which occur when using oxy-acetylene welding and cutting equipment result from the lighting of a cigarette in the presence of an oxygen enriched atmosphere. Therefore it is impossible to over-emphasise the danger of smoking in oxygen enriched atmospheres or where oxygen enrichment can occur. In such areas smoking must be forbidden and consideration given to prohibiting the carrying of smoking materials

The second factor is the speed at which flammable vapours are released from the fuels present. This will depend on the type of material, its size and the temperature it is being exposed to. Some materials, such as flammable liquids will readily give off vapours and will significantly increase the amount of vapours given off as the temperature is raised by a small amount. Other materials, such as a large block of wood, may not readily give off vapours until exposed to particularly high temperatures. The size of the flammable material is significant in that it is small and finely divided, such as a dust or droplet of fuel. The rate of vapour being evolved and fire growth will be very rapid and may resemble an explosion.

How fire spreads

CONDUCTION

Conduction may occur in solids, liquids or gases although it is most clearly present in solids. Examples of the effect are where a teaspoon in hot tea transfers heat to the hand of the person holding it. Most metals are good conductors, silver and copper being the best. Glass, wood, cork and asbestos are poor conductors.

In fire situations a steel girder passing through an otherwise fire resistant wall may cause fire spread to another part of a building by conduction. A steel door subjected to heat on one side conducts heat rapidly to the other side, but a wooden door (although it may ignite) is, initially, a more effective barrier to heat due to its being a poor conductor.

This method of fire spread is prevented by insulating structural steelwork with fire resistant materials so they do not gain heat and conduct it to another area. This was originally done with asbestos, but is now done with dry linings, and spray coatings.

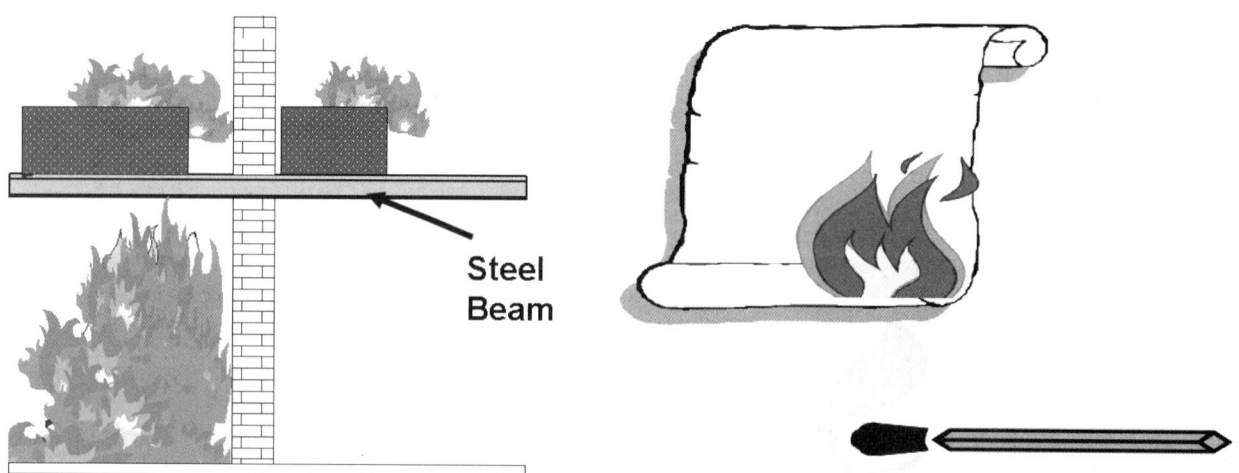

Figure 2-9: Conduction.   *Source: FST.*   Figure 2-10: Direct burning.   *Source: FST.*

CONVECTION

Convection only occurs in liquids and gases. When a liquid or gas is heated it expands and therefore becomes less dense. The lighter liquid or gas rises, being displaced by colder and therefore denser liquid or gas. The cooler liquid or gas in turn becomes heated and so a circulation is set up.

Convection is used in domestic hot water systems and in heating systems using so-called 'radiators'. Most of the heat from these radiators is in fact carried away by convection. Convection also causes the updraft in chimneys. In a fire situation in a building, convection currents can carry hot gases produced by combustion upwards through stairwells and open shafts, thereby spreading fire to the upper parts of buildings.

A typical fire may start in a waste paper bin. As the bin burns the heat given off from the burning materials will rise. The heat layer produced may be sufficient to cause the surface linings on the ceiling to ignite. This would then cause an increase in the size of fire, with a resultant increase in heat output. The hot gases produced will in turn spread laterally across the ceiling and will rise higher throughout the building.

This method of fire spread is prevented by fire separation and compartmentation of buildings by the use of fire resisting doors / walls, and fire stopping of openings.

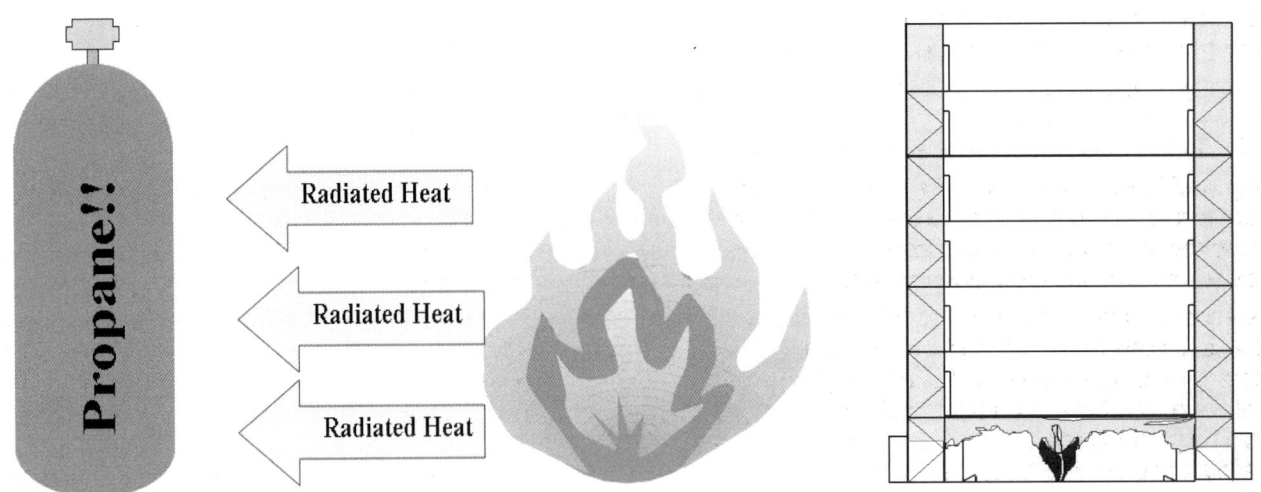

Figure 2-11: Radiation.   *Source: FST.*   Figure 2-12: Convection.   *Source: FST.*

RADIATION

Heat may also be transmitted in straight lines by a means which is neither conduction nor convection, for example the process by which heat from the sun passes through empty space to warm the earth. This method of heat transmission is named 'Radiation' (infrared radiation) and does not involve any contact between bodies and is independent of any material in the intervening space.

Many fires have been caused by radiation - one of the most common is clothing being ignited whilst being dried near to a gas fire used to heat a room, electric heater or a boiler in the workplace. Radiant heat from the suns rays passing through a glass window have been known to cause fires by the rays being concentrated through a lens type object such as a glass paper weight onto combustible material. If a fire were to start in a warehouse the burning materials might radiate sufficient heat energy to ignite the next stack of materials positioned nearby.

This method of fire spread is prevented by physical space separation between buildings and/or stocks. If space separation cannot be achieved, then a fire resistant barrier may need to be erected.

DIRECT BURNING

One of the main methods of fire spread is by physical flame contact. As a material burns, it has the potential for the flames from the combustion process to touch and ignite other materials nearby.

COMMON BUILDING MATERIALS

General considerations

The properties of building material vary considerably and this will affect their choice in relation to fire hazards. The choice of materials used may also be affected by other parameters such as economics, availability and aesthetics. The principal consideration should be to ensure that the material and its application comply with the law, local regulations, British and European standards. All building works, with very few exceptions, are controlled by some form of building legislation. The principal aim of this legislation is that materials are used correctly so that the safety of life is ensured, in the event of fire.

When considering the type of material to be used, and its application, the following criteria should be assessed:
- Ignitability.
- Flammability.
- Surface spread of flame.

All the above relate to the safety of the occupants of a building, as they may affect the means of escape, whilst the following are probably more important from the point of view of the safety of the building (damage to the structure and contents):

- Heat release.
- Smoke (or gas) release.
- Fire resistance.
- Flame penetration.
- Smoke (or gas) penetration.

Other considerations that may affect the choice of building materials include:

a) The use or uses of the building.

b) The dimensions (compartmentation) of the building.

c) The design and layout, including escape routes.

d) Whether or not the material will burn.

e) The ability of the material to support and spread flame across its surface.

f) The behaviour of the material when it is burning.

g) The effects of high temperature on non-combustible materials.

The effects of fire on building materials will vary according to its application to the structure.   Consideration must be given to the effects of materials being combined in use.   The reaction of a specific material to fire is likely to be entirely different if that material is used in a different format, applied differently, or used in conjunction with other materials.

Timber - factors affecting fire resistance

The fire resistance of timber as an element of structure depends upon the following:

- The density of the timber.
- Its thickness and cross sectional area.
- The quality of workmanship and/or detailing.

Timber has a very low thermal conductivity and this factor, combined with its production of a protective skin of charcoal in a fire, retards its rate of combustion.   Carefully constructed joints which eliminate cracks and air spaces contribute substantially to fire resistance.   Due to this slow rate of burning, and the fact that timber will distort and sag for a considerable way before collapse occurs, it is a fairly good material to use when considering fire in buildings.

Timber can be treated with fire retardant materials to increase its safety within fires.   Fire retardant coatings can be applied to timber with a class 3 or class 4 rating for surface spread of flame to raise its classification to class 1.

Bricks - factors affecting fire resistance

Brickwork can provide a vital role as a non combustible and fire resistant element as part of the structure of a building, and can also be used to protect other elements of structure from the effects of fire, e.g. loadbearing columns, staircases, shafts and ducts.

Loadbearing brickwork has inherent fire resistance and requires no further protection.   The duration of fire resistance is dependent upon the thickness of the wall: i.e.

- A 100 mm brick wall gives approx. 2 hours fire resistance.
- A 200 mm brick wall gives approx. 4 hours fire resistance.

The fire resistance is also affected by the type of construction of the brick itself. If the brick is hollow cast or has holes throughout its width, then it is more susceptible to spalling with the bottom face falling off.   This would obviously reduce the fire resistance.

Concrete - factors affecting fire resistance

Concrete is strong in compression but weak in tension and is therefore reinforced with steel in areas which will be subject to tensile stress, e.g. lower half of a concrete beam.

The fire resistance of concrete elements is influenced by the following:

a) Size and shape of element.

b) Disposition and properties of reinforcement.

c) The load supported.

d) The type of concrete and aggregate.

e) Protective concrete cover provided to reinforcements.

f) Conditions of end supports.

As stated previously, concrete relies heavily on reinforcement steels for its tensile strength.   If the steel reinforcement is allowed to heat up then it begins to lose strength.   There is a temperature at which concrete reinforcement materials will lose 50% of their strength. This point is known as the 'critical temperature'.   The critical temperature for mild steel reinforcements is $550^{\circ}C$ and that for high yield steel reinforcements is $400^{\circ}C$.

Metals - factors affecting fire resistance

Metals may require surface protection to minimise the danger of fire spreading by conduction. Unprotected metal used structurally may also present a danger of collapse in fire. All metals soften and melt at high temperature. Structural steel loses 2/3rd of its strength at 600°C and begins to sag and distort.

Metals expand when they get hot. A steel joist 10m long will expand by 8 cm when heated to 600°C. This expansion factor, especially as the length of the beam increases, can cause walls to be pushed out with a resultant structural collapse.

Aluminium alloys are now being used in building construction because of their unique properties, as follows:

a) The reduction in weight (aluminium is 1/3rd the weight of steel).

b) Resistance to corrosion.

c) The ease of handling and working.

d) High strength to weight ratio.

However these types of alloys have the following disadvantages:

a) Very rapid loss of strength in fire (stability affected at 100°C - 225°C).

b) High expansion rate (twice that of steel).

c) High thermal conductivity (three times that of steel) giving a greater risk of fire spread.

d) Low melting points (pure aluminium 658°C).

Steel or metal alloy structural members must therefore be protected from the effects of fire. This can be done in one of the following ways:

a) Solid protection.

b) Hollow protection.

c) Sprayed or applied mineral coatings.

d) Intumescent coatings.

e) Hollow section filled with water.

f) By filling hollow webs of beams etc. with lightweight blocks of concrete.

g) By design features such as suspended ceilings.

Building boards and slabs - factors affecting fire resistance

Fire resistance and surface spread of flame characteristics are inherent qualities of board materials. If the performance qualities of boards are low then it may be necessary to increase this by the addition or impregnation of a fire resistance substance either onto or into the board material.

Escape routes and circulation spaces within buildings should have both ceilings and walls comprising materials of class '0' standard.

The surface spread of flame characteristics gives an indication of the speed that fire would spread across a materials surface. Class 1 would be the slowest speed with class 4 the fastest speed. Class 0 is not a true classification, but to be categorised as class 0 a material must be class 1 and must not contribute greatly to the propagation of the fire.

Different materials are given differing classification, some examples of which are given below:

- Plasterboard   Class '0'
- Woodwool slabs   Class '0'
- Mineral fibre board Class '0'
- Chipboard   Class '3'
- Softboard   Class '4'
- Plywood   Class '3'

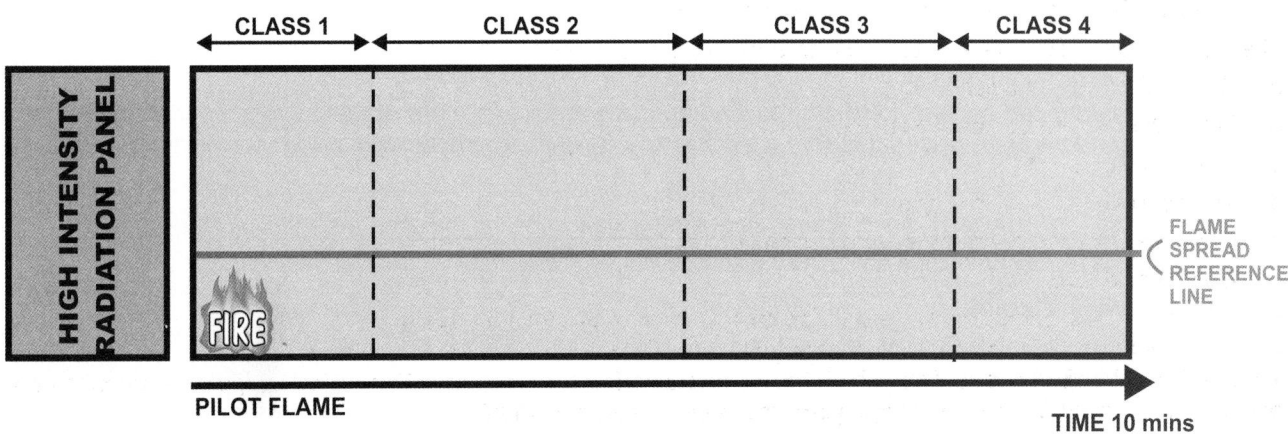

Figure 2-13: Horizontal spread time chart.

Source: Dulux Pyroshield.

Sandwich panels

In modern buildings one type of construction which is commonly used, but is causing great concern, is known as 'sandwich panels'. This fabrication consists of two outer skins of sheet metal (normally a light alloy) with an infill of heat insulating material. In some cases the material used is polyurethane or styrene foam. Buildings constructed from sandwich panels are liable to sudden, unpredictable collapse when a fire occurs, accelerated by the panels falling out of their framework. Very rapid fire spread occurs as the heated / exposed foam breaks down into volatile flammable toxic gases.

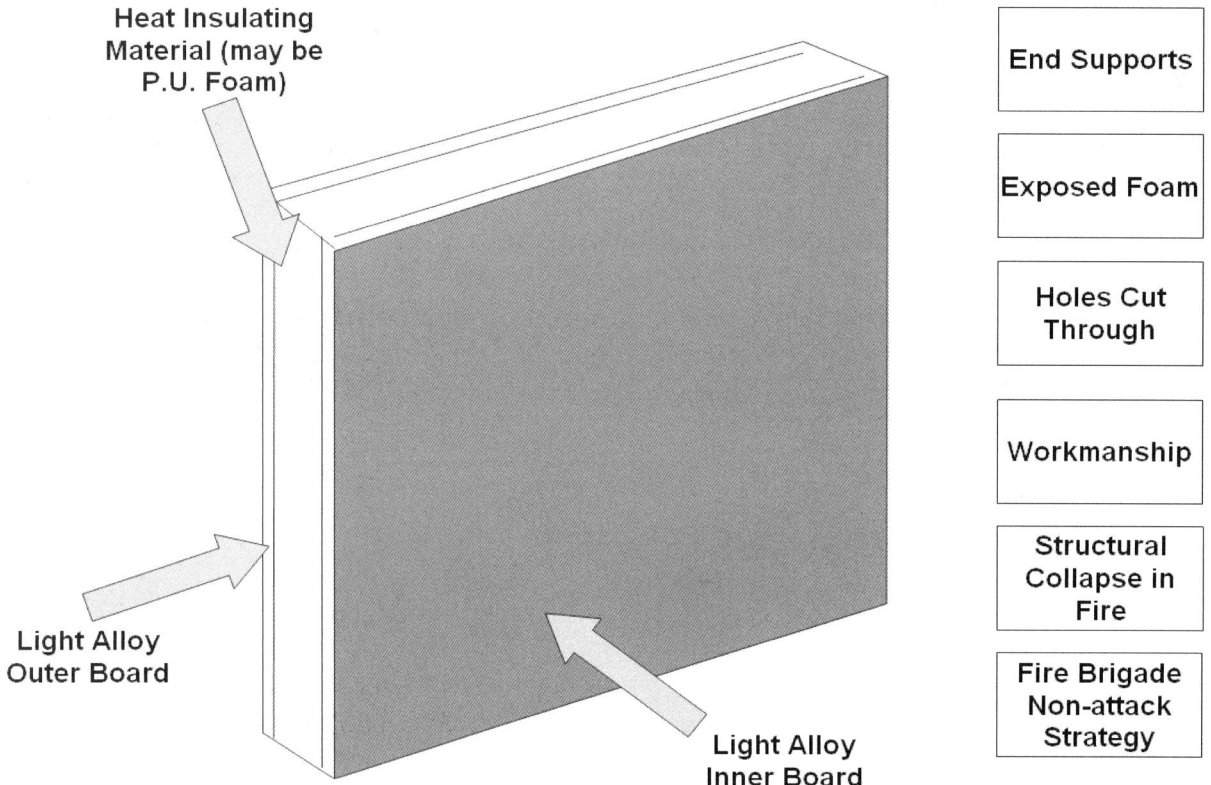

Figure 2-14: Sandwich panels.   *Source: FST.*

Each fire of this type will have to be assessed for risk and fire brigades may have to consider attacking a fire from outside the building, so as not to endanger the lives of fire-fighters.

Plastics - factors affecting fire resistance

Basically there are two types of plastic:

1. Thermoplastics - which when heated will soften and melt.

2. Thermosetting - sets to a hard infusible form.

Fire resistance

Plastic materials are composed of combustible organic material and will have limited resistance to fire and to fire spread. Most plastics when exposed to fire emit a considerable amount of smoke and toxic fumes.

Glazing materials - factors affecting fire resistance

Glass is a non-combustible material and therefore will not contribute to the fire load (the amount of combustible material) of a building. Standard glass panels in doors or walls create a weak point in fire compartmentalisation caused through early collapse or their transparency to radiated heat enabling fire to spread.

Fire resisting glazing can be used to give up to 1½ hours' fire resistance. The actual figure is dependant upon the nature and dimensions of the glass panel, the type of frame and the glazing detail such as the seal method to the frame.

Wired

This can be used to give up to 1½ hours' fire resistance dependant upon the materials and design of the frame.

The glass used is usually 6mm thick and up to 1.6 square metres in area.

Laminated glass (Pyran)

This type of glass is now becoming more common in use. It comprises 3 to 5 layers of glass with interlayers of intumescent material which reacts at 120°C to form an opaque shield. This glass composite prevents radiated heat passing through and can therefore be used in many more situations than the traditional wired glass.

The nature of an open burning fire

In an open burning fire outside a building, the hot gases of combustion will rise into the atmosphere with very little effect on the temperature of the materials involved in the fire. As a result the speed of fire growth is generally much slower than it would be in an enclosed space.

Fire in a physical enclosure

A fire in an enclosed room is however a totally different phenomena to an open burning fire. The speed of fire growth can be devastating, with two fire phenomena of specific note - flashover and backdraught.

FLASHOVER

A flashover can occur when a fire is free burning in a room. For this to happen there must be a good supply of air either from the large dimensions of the room, an open door, open window or ventilation system. As the item that was initially ignited burns and the fire grows, the radiated heat heats up all other materials in the room until they reach their spontaneous ignition temperature. Items in the room will then instantly ignite, creating the impression that the fire has flashed from one side of the room to the other. Although this phenomena is very serious, it is not the most dangerous, as it is obvious that there is a severe fire in progress.

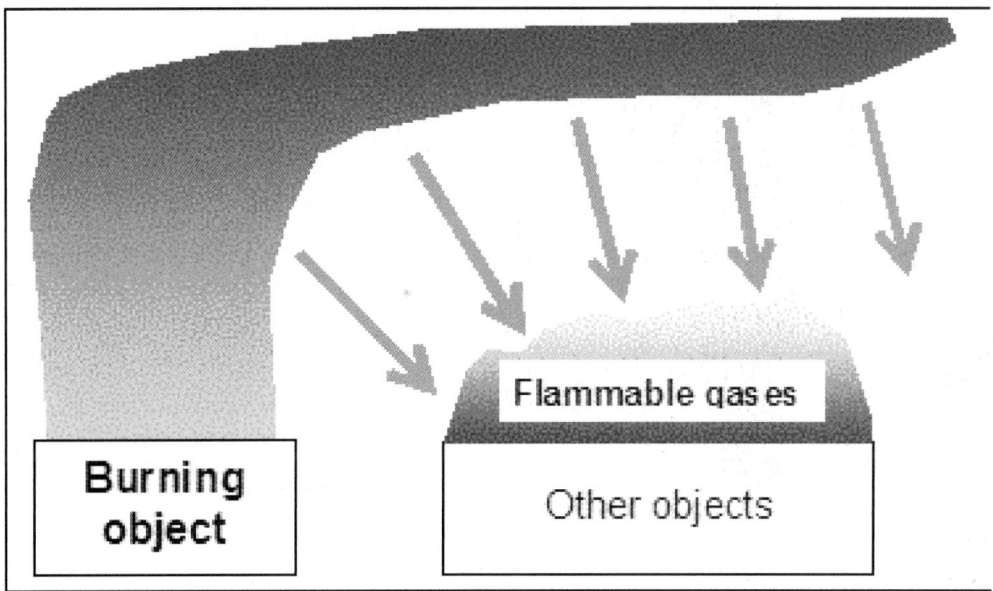

Figure 2-15: Flashover.   *Source: FST.*

BACKDRAUGHT

The most dangerous fire phenomena is known as a 'backdraught'. For backdraught to occur a fire must start in a closed room environment - for example where windows and doors are shut, with little air flow into the room. As the initial fire burns it will use up the oxygen within the room. The fire can then die down, but often does not go out. The smouldering fire that remains will then fill the room with large quantities of high temperature smoke. As smoke consists of flammable gases mixed with combustible fuel particles it would normally burn off. However, due to the lack of oxygen in the room these flammable products will not burn, even though they are at a temperature and above their spontaneous ignition temperature. When someone opens the door to the room and lets oxygen into this environment the flammable gases/fuel can instantly ignite. Should this happen, it does so with an explosive force that will drive the flames back out of the opening that the air came in from. The flames that are produced will come out of the opening at 10 metres per second and 1100°C (30 feet per second and 2000°F). Backdraught is the most dangerous fire phenomena, and the risk associated with opening doors onto a possible fire situation should be included in all employee fire training.

Fire growth rates

The fire growth rate is mainly affected by the rate of production of flammable fuel in the form of vapours and the availability of oxygen.

SMOKE MOVEMENT

Smoke movement will be dictated by the temperature of the smoke. Cold smoke will spread laterally possibly at low level, presenting a principle hazard of reduced visibility. Whereas hot smoke is more hazardous, since it is a fuel above its ignition temperature and will spontaneously combust when it reaches available oxygen. Hot smoke will be carried with the convection currents produced by the combustion process of the fire and will spread laterally across the ceiling then upwards at every opportunity. Due to the buoyant nature of hot smoke it can travel a considerable distance from the seat of the fire.

It is important to consider the make up of smoke. Smoke consists of two basic components:

- The particles within the smoke which are in fact unburnt carbonaceous material.
- The fire gases which are mixed in with the smoke - these gases are both flammable and toxic.

The majority of smoke will contain carbon monoxide (proportional to the available oxygen) along with other toxins such as hydrochloric acid from the combustion of electrical conductor insulation, and compounds of cyanide from combustion of synthetic upholstery foam infill. Such gases will quickly render people unconscious and cause death and helps to explain why people who die from fires that ocur whilst they are asleep tend not to attempt to escape.

THE EFFECT OF BUILDING CONSTRUCTION

The style and method of construction can play a major part in fire growth, as can the maintenance of structural elements. If a building is constructed with large open areas with little or no fire resistance or compartmentation, the probability is that a fire will grow easily unless it is suppressed by some extinguishing method. Large open plan and open floor buildings do very little to limit the free air that can feed a growing fire and flashover is a real possibility. It is important to check the construction of a building to see what easy paths are available for smoke and fire to circulate around the building. Areas of concern may be above false ceilings and behind wall panelling as building work may have gone ahead without the proper control measures in place. Other areas of concern would be:

- Vertical shafts such as lifts.
- Open stairways.
- False ceilings.
- Voids behind wall panelling.
- Doors which are ill fitting, damaged or wedged open.

Any holes in fire resistant structures should be 'fire stopped' to maintain the integrity of the construction. The materials used in the construction are also vital to the rate of growth of a fire within the building. If the construction materials are non combustible, the fire will only grow as a result of the contents of the building. If the construction materials are principally combustible, for example timber, the speed of fire growth will potentially increase. However, the rate of growth of fire can sometimes be affected by other circumstances, for example the level of insulation built into the building construction. If a room's walls are lined with non-combustible insulating material the time taken for a fire to reach flashover stage is dramatically reduced.

THE EFFECT OF VENTILATION

Ventilation is the key factor in fire growth. If a fire does not have sufficient oxygen to burn properly then it will extinguish with time. If a fire has unlimited ventilation / air supply then a flashover can easily occur. If ventilation to a fire within a room is limited, then there is potential for a backdraught to occur. Ventilation systems will contribute to fire spread in two ways. Firstly, a ventilation system can provide a supply of fresh air (oxygen) to feed a fire and secondly smoke (which is a fuel) can cause fire to spread through ventilation systems to other parts of the building, and bypass fire / smoke resistant structures.

THE EFFECT OF CONTENTS

When considering the possibilities and effects of fire spread within a building due to the types of materials that are stored and used within it, HSG 64 "Assessment of Fire Hazards from Solid Materials" gives guidance on this matter. This document subdivides materials into two categories, *high* or *normal* risk. This classification is based on two different tests related to combustion properties. The first test specifies materials high or normal risk dependent upon the maximum rate of temperature rise, the second test specifies the risk on the volume of smoke produced by the material. As can be expected some materials will be classified as high in one test but normal risk in another. These tests are used beacause the rate of temperature rise and the amount of smoke being produced from a material will have a major influence on people's immediate ability to use the means of escape provided. They are therefore a useful test in helping to make a risk assessment of the risk to life from the use of a given material.

Examples of materials which fall into the high risk category in both tests, and therefore present a great danger include:

- Acrylic fibre.
- Acrylic mixture.
- Acrylic over locks.
- Expanded polystyrene.
- Flexible polyether (Poly Urethane foam).
- Polypropylene sliver.
- Rigid Poly Urethane foam (low density).

Other considerations include, type, quantity, distribution and reactive nature of materials stored.

2.4 - Explosion

Description of the mechanism of explosion

Explosions can occur in flammable gases/vapours and also in certain types of dusts. An explosion is probably best defined as "rapid flame propagation throughout an area containing flammable gases, vapours and dusts". For the explosion to occur the gas/vapour/dust must be mixed with air in such proportions that the mixture is within the flammability range for that substance. Explosion can occur with such gases as hydrogen, propane, acetylene and examples of dusts that may cause explosion hazards are aluminium, coal, flour and polythene. In a dust explosion, there is an initial smaller 'Primary Explosion', which is then followed by a devastating 'Secondary Explosion'.

PRIMARY AND SECONDARY EXPLOSION

Industrial dust explosions can be divided into two types. Firstly a primary explosion, which usually occurs in an enclosure or powder handling plant located within a building. Structural damage of lightweight plant may occur at this stage, as pressures of 8 - 10 bars are produced, but more concerning is the consequent air turbulence created by the primary explosion within the building. The pressure wave created combined with the air turbulence may dislodge accumulated dust from all horizontal surfaces within the affected parts of the building, and cause an airborne suspension of combustible dust throughout the affected area. Dislodged dust can then be ignited by either the initial ignition source, the combustion of products of the primary explosion or any other ignition source with sufficient heat energy - causing a secondary explosion. The secondary explosion will then travel throughout the entire workplace with devastating effects. Entire buildings have been destroyed by such effects. Wherever possible equipment should be designed to contain the primary explosion or to vent to a safe area.

The types of industries that are typically at risk are those dealing with agricultural products, foodstuffs, pharmaceuticals, chemicals, pigments, polymeric materials, rubbers, coal and wood products.

Definitions

DEFLAGRATION

Deflagration is a process of subsonic combustion that usually propagates through thermal conductivity. This is where hot burning material heats the next layer of cold material and subsequently ignites it. This then continues very rapidly with layer upon layer of material.

A deflagration occurs when a flammable gas / vapour becomes mixed with air to form a flammable gas / air mixture and is exposed to an ignition source. A shock wave may be produced if the explosive force is contained. This can propel heavy objects through the air causing structural damage and injury or death to people. If a deflagration occurs inside a building the pressure waves that are created can cause considerable structural damage as they can travel at sub-sonic speeds. Typical materials that have a tendency to deflagrate are propane, butane and methane.

DETONATION

A detonation is the most devastating form of gas explosion. Unlike deflagration, a detonation does not require confinement or obstructions in order to propagate at high velocity. The behaviour of a detonation is quite different from a deflagration, particularly in an unconfined situation. A detonation is defined as a supersonic combustion wave. In fuel / air mixtures at atmospheric pressure, the detonation velocity is typically 1500 - 2000 metres per second (m/s) and the peak pressure is 15 - 20 bar pressure.

The probability of an occurrence of a detonation in fuel / air mixtures depends strongly upon the type of fuel. Very reactive fuels, such as hydrogen, acetylene or ethylene may detonate in an accident situation. In accident situations involving such fuels detonations should be regarded as a possible scenario.

UPPER EXPLOSIVE LIMITS

A flammable mixture will only explode in air if the mixture lies between certain limits. The 'Upper Explosive Limit' (UEL) is the highest mixture of fuel and air that will just support an explosion.

They are normally given as a percentage of the gas/vapour in air or weight of the dust in a volume of air (Kg/m3).

LOWER EXPLOSIVE LIMITS

The 'Lower Explosive Limit' (LEL) is the lowest mixture of fuel and air that will just support an explosion.

The conditions for gas and dust explosions to occur

Explosions can only occur if certain conditions are present:
- The gas / dust must be combustible.
- The gas / dust must be capable of becoming airborne and mixed with the air.
- The dust particle size and distribution must be capable of propagating flame.
- The concentration of gas / dust must fall within the explosive range.
- An ignition source of sufficient heat energy must be in contact with the dust.
- The atmosphere must contain sufficient oxygen to sustain combustion.

The principles of explosion suppression

METHODS OF EXPLOSION RELIEF

Explosion venting panels

'Explosion Relief Venting' is the most common system used and involves incorporating deliberate points of weakness in a structure. These normally take the form of explosion relief vents in the process plant and/or building. If the vents are of the correct size and in the correct place an explosion will be vented to outside.

The objective is to prevent the explosion pressure from exceeding the design strength of the plant or building. This is normally done by use of lightweight roofs, lightweight wall panels, louvres and vents.

The size of vents required will depend upon the properties of the dust involved, the strength of the plant/building involved and the opening pressure of the vents.

Vent panels may become dangerous missiles in the event of an explosion, so they may need to be fixed to the plant or building with chains or some similar device.

Care needs to be taken with the siting of vents in order to prevent the fireball or pressure wave that is produced from creating further dangers.

Bursting discs

Certain plant and machinery will have explosion relief built into it in the form of a bursting disc. This is a purpose designed weak spot that is designed to rupture at a pre-determined pressure. This will therefore vent any pressure immediately and will prevent a more damaging explosion occurring.

Suppression (e.g. inerting)

It may not be possible or desirable to provide explosion relief venting, and a method of explosion containment or suppression may need to be installed.

An explosion may develop pressures up to 10 bars, a pressure which buildings cannot withstand, but small items of plant such as grinding equipment may be able to be designed to do so. However, this method is not often cost-effective on larger items of plant/machinery.

An explosion suppression system will detect an explosion in its early stages by detecting a pressure increase. The system will then inject an extinguishing agent (often dry powder) into the path of the explosion so that the flame is extinguished, and the explosion does not proceed.

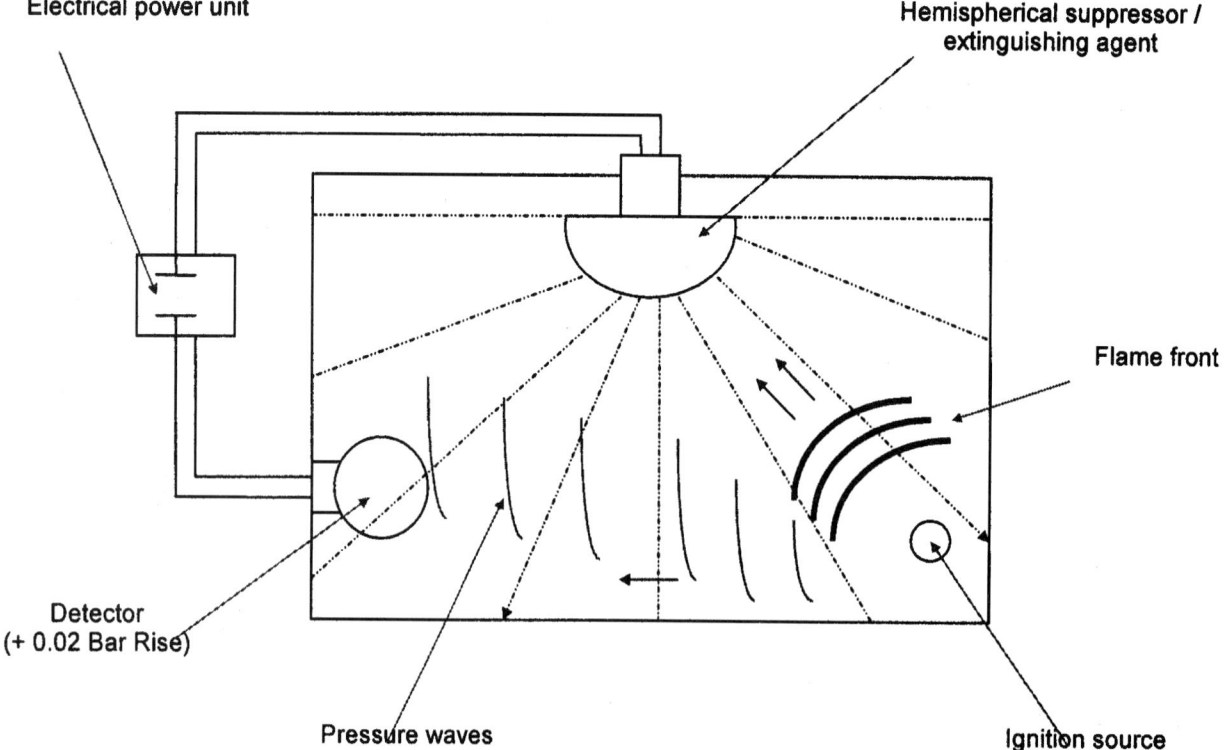

Figure 2-16: Explosion suppression.

Source: FST.

Causes and prevention of fires

Overall Aims

On completion of this Element, candidates will understand:

■ common causes of major fires.

■ fire risks associated with flammable, combustible and explosive substances.

■ fire risks associated with common workplace processes and activities.

■ measures to prevent fires.

Content

Specific Intended Learning Outcomes

The intended learning outcomes of this Element are that candidates will be able to:

3.1 identify the causes of fires and explosions in typical work activities

3.2 outline appropriate control measures to minimise fire risks

3.3 advise on the requirements for ensuring that flammable storage facilities meet current regulatory and best practice standards and guidelines

3.4 identify and minimise risks from arson

Relevant Statutory Provisions

Regulatory Reform (Fire Safety) Order (RRFSO) 2005

Dangerous Substances and Explosive Atmospheres Regulations (DSEAR) 2002

Chemicals (Hazard Information and Packaging for Supply) Regulations (CHIP) 2002

Control of Major Accident Hazards Regulations (COMAH) 1999

3.1 - Causes of fires

When the Fire and Rescue Service attends a fire in any workplace, they have to complete a report on the fire. These reports build into a national picture when they are compiled annually. The statistics in this note are taken from the National Fire Statistics 2004 (with statistics on house fires removed).

Causes of fire can be divided into 2 main groups - accidental and arson.

Accidental fires are considered below, arson is considered later in this element.

ACCIDENTAL FIRES

Accidental fires account for 57% of all fires in the UK. This group can be divided into 3 sub groups:

- Careless actions - 26%.
- Misuse of equipment and appliances - 24%.
- Defective equipment - 50% (Where faulty appliances and leads accounted for 78% of these).

Careless actions

As can be seen careless actions are also a major factor in accidental fires and therefore must be addressed. Carelessness is attributable to people, and we must therefore inform, educate, train and, if necessary, control people to prevent fires being caused in this way.

Misuse of equipment and appliances

Figure 3-1: Misuse of equipment.   *Source: ACT.*

Figure 3-2: Misuse of equipment.   *Source: FST.*

People misuse equipment in two ways, inadvertent misuse and deliberate misuse. For example, if you push a folder down the side of a computer blocking the air vent, you have inadvertently misused the computer. If you overrate the fuse to the computer on purpose, then you have deliberately misused it. In both cases it is people who carry out the acts of misuse. The Regulatory Reform (Fire Safety) Order (RRFSO) 2005 requires the Responsible Person to ensure that their employees are provided with adequate safety training covering appropriate precautions and actions in order to safeguard the employee and others on the premises.

Defective equipment

Defective equipment and the resultant fault may lead to localised heat around the area of the fault and over time this heat may build up and lead to ignition of materials located near by. To prevent such fires it is important to look critically at the preventative maintenance schedule for equipment within the workplace. If the maintenance programme for equipment were to be reduced, for cost saving reasons, there is a risk that it will be reduced too far.

IDENTIFYING SOURCES OF IGNITION

The majority of fires need an ignition source to enable them to start. It is imperative therefore that an assessment is made at workplaces to recognise all ignition sources that are, or are likely to be, present. Once these ignition sources have been located, then the surrounding areas can be checked for combustible materials or gases that are normally, or are likely to be, present.

Having now identified all problem areas, a study can be made to assess the viability of removing the ignition source and the combustible material from each other. If this is not possible then some form of control measure should be taken so as to minimise the risk of fire.

It is also important that good housekeeping is maintained. If any of the above ignition sources are present and the housekeeping is poor, then the likelihood of a fire breaking out is dramatically increased.

MATERIALS IGNITED

If we look at the statistics overall for both accidental fires and arson, we get the following results:

Material First Ignited	% of Total	% of Fatalities	% of Injuries
Paper, Cardboard	18%	Nil	16%
Electrical Insulation	16%	4%	11%
Food	11%	2%	13%
Structure & Fittings	13%	Nil	4%
Textiles, Upholstery & Furnishings	8%	18%	17%
Flammable Liquids	5%	13%	12%

Figure 3-3: Accidental fires and arson statistics.   *Source: ACT.*

Under the RRFSO there is a legal duty to prevent fire. Therefore we should consider these statistics when undertaking fire risk assessments.

Common sources of ignition of major accidental fires

Ignition sources can be found in many forms and some of the more common examples are listed below:

- Smokers' materials.
- Sparks from welding equipment.
- Sparks from electrical motors.
- Sparks from grinding equipment.
- Oxy-acetylene welding equipment.
- Fixed or portable heaters.
- Cooking equipment.
- Bitumen boilers.

- Electrical faults.
- Overloaded electrical circuits.
- Overheating equipment.
- Static electricity.
- Non-intrinsically safe equipment used in a flammable atmosphere.
- Radiated heat from a legitimate source such as a light bulb.
- Hot surfaces such as soldering irons or hot glue guns.
- Steam pipes.

ELECTRICAL APPLIANCES AND INSTALLATIONS

Examples of defects, acts or omissions that may cause fires arising from electrical equipment are listed below:

- Electrical short circuits, insulation failure or earth faults.
- Blocked air vents.
- Appliances or machinery which is not intrinsically safe being used in a flammable atmosphere.
- Poor electrical connections, allowing arcing and resistance heating.
- Overloading electrical circuits by up-rating fuses to carry more electrical current.
- Electrical equipment left switched on when not in use (unless it is designed to be permanently connected).

LIGHTNING

There is always the potential for a lightning strike on a building, especially any tall building. A lightning strike can cause a power surge and localised heating to cables and/or equipment. Lightning strikes would be a potential source of ignition for large, tall chemical process plant or for tank farms that contain flammable liquids. This would particularly be the case where they are set in a rural location where they may present the highest point to any storm containing lightning.

Lightning protection systems should be installed where buildings are at significant risk and be subject to periodic inspection and test. The typical and simplest form of lightning protection would be a lightning conductor, which is a heavy duty conductor which is run from above the highest point of the building to earth. It is designed and positioned such that any lightning will tend to strike the lightning conductor in preference to the building, allowing the energy to flow to earth via the conductor.

COOKING

Many businesses have kitchens where staff may prepare food themselves. These facilities are similar to domestic kitchens and similar cooking hazards will exist.

Major causes of kitchen fires include:

- Food left unattended on the cooker.
- Clothes left to dry over a heat source.
- Faulty electrical appliances.
- Electrical appliances left on and unattended.
- Spillages of flammable oils near to heat sources.
- Deep fat frying, unless a thermostatically controlled pan is provided. Such equipment should not be left unattended.

- Combustible materials such as cloths, towels and loose fitting clothing (especially sleeves) not kept well clear of hobs.
- Toasters and microwaves left unattended / poorly sited.

HEATING AND LIGHTING

A number of fires are caused by carelessness or misuse of heaters and lighting units. Some of the main causes of fires in this way would be:

- Placing portable heaters too close to combustible materials.
- Covering the air vents on top of heaters, particularly with combustible materials.
- Combustible materials placed too near electric lights, for example, in a warehouse
- Electrical 'task lighting' placed too near combustible materials, for example, in an office, workshop or construction site. Particularly where the lighting is high wattage, for example, the halogen type of lights used in temporary workplaces. An example of a fire caused by this source of ignition would be the Windsor Castle fire in November 1992 when a curtain came into constant contact with a temporarily installed halogen light.
- Fixed appliances sited too near to combustibles.
- Non-intrinsically safe lighting used in a flammable atmosphere.

 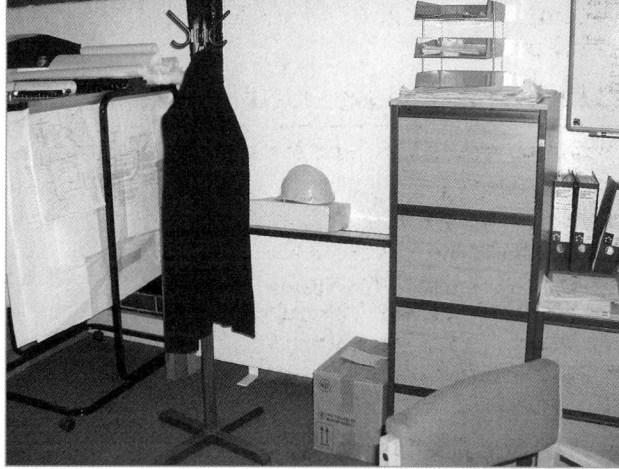

Figure 3-4: Poorly sited heater.   *Source: FST.*   Figure 3-5: Combustible material on top of heater.   *Source: FST.*

SMOKING

Careless disposal of smoking materials has historically been a frequent cause of fires and fire deaths. In recent times, smoking materials have caused 7% of all accidental fires in non residential premises. Fires caused in this way should reduce as in many cases those responsible for buildings prohibit smoking. However, any smoking prohibition should be thought out fully as it may lead to illicit smoking in areas that are not well controlled or monitored. Prohibition is essential where flammable materials are stored and processed, or in situations where ignitable / explosive dusts, vapours or readily combustible waste is produced.

Smoking should be prohibited in stock rooms and other rooms not under continuous supervision. Any area where 'no smoking' is imposed should have the rule strictly enforced.

Where smoking is allowed easily accessible, non-combustible receptacles for cigarette ends and other smoking material must be provide. The area must be under the control of a responsible person, containers emptied daily and the area checked regularly or at close of the day. Smoking should cease half an hour before closing down to reduce the risk of a fire occurring that does not get observed till it is out of control.

Fire and explosion risks from flammable materials

GENERAL SAFETY PRINCIPLES

The main hazards associated with flammable materials are fire and explosion, involving the product itself or any vapours it may produce. Fires and explosions are likely to occur when dusts, vapours or liquids are released from a controlled environment into an area where there may be an ignition source. The reverse is also true, where an uncontrolled ignition source is introduced into an area where any substance that can burn is present.

IN USE WITHIN THE WORKPLACE

Common causes of fire and explosion in areas where flammable materials are being used in the workplace are:

- Lack of awareness of the properties of flammable materials.
- Operator error, due to lack of training in safe use, handling and storage of flammable materials.
- Hot work on or close to flammable material containers.
- Inadequate design of equipment.
- Failure or malfunction of equipment.
- Exposure of flammable materials to heat from a nearby fire.
- Misuse of flammable liquids, for example, to assist in burning waste materials.
- Inadequate control of ignition sources.
- Electrostatic discharges.
- Heating materials above their auto-ignition temperature.
- Dismantling or disposing of equipment containing flammable materials.

Figure 3-6: Unsuitable container for transport.   *Source: FST.*

Figure 3-7: Unsuitable container for use.   *Source: FST.*

IN STORAGE WITHIN THE WORKPLACE

Hazards and risks created by the storage of flammable materials in the workplace are very similar to those related to use in the workplace. Common causes of fires and explosions in and around storage facilities within the workplace are:

- Lack of awareness of the properties of flammable materials.
- Operator error, due to lack of training.
- Inadequate or poor storage facilities.
- Hot work on or close to flammable material containers.
- Inadequate design, installation or maintenance of equipment.
- Decanting flammable materials in unsuitable storage areas.
- Exposure to heat from a nearby fire.
- Inadequate control of ignition sources.
- Dismantling or disposing of containers containing flammable materials.

Figure 3-8: Deliberate ignition of materials.   *Source: ACT.*

Figure 3-9: Not knowing the sound of an alarm.   *Source: ACT.*

General points to prevent fires involving flammable materials in storage within the workplace:

- Quantities of flammable materials in storage should be minimal, with sufficient stocks to efficiently run the business, but without overstocking.
- Under traditional guidance you can store up to 50 litres of highly flammable liquids in the workplace in a fire resisting container.
- External storage is preferred to internal storage. If a fire wall is used it should normally provide a fire resistance for a minimum of 30 minutes.
- Storage area should be well ventilated.
- Where possible, containers of flammable liquids should be stored in a bund, drip tray or provided with other spill containment.
- Empty highly flammable liquids and liquefied petroleum gas containers should be stored separately from full ones.
- Prevent container from external damage, for example, impact damage from fork lift trucks.
- Isolate damaged or leaking containers.
- Reduce fire and explosion risk by creating separation between storage containers and limiting the height of stacks.
- Lighting for storage areas for flammable liquids must be 'flameproof' or similar.
- The storage facility needs to be secured from unauthorised access.
- Hazard warning signs should be displayed.
- Any persons storing flammable material should be trained.

IN TRANSPORT WITHIN THE WORKPLACE

Incidents involving flammable materials commonly occur during transport or transfer operations, including:

- Movement from storage.
- Decanting or dispensing.
- Movement within premises.
- Unsecured storage within vehicle.
- Emptying vehicle fuel tanks.
- Dealing with spillages.

Common causes of fires and explosions relating to transporting flammable materials within the workplace are:

- Lack of awareness of the properties of flammable materials.
- Operator error, due to lack of training.
- Inadequate or poor transport facilities, for example, transporting flammable liquids in open buckets.
- Inadequate design, installation or maintenance of equipment, for example, equipment used to pipe flammable liquids.
- Absence of spill strategies.
- Hot work on or close to flammable material containers.
- Electrostatic discharge.

Fire risks in construction and maintenance work

SITE STORAGE OF COMBUSTIBLE AND FLAMMABLE MATERIALS

Many serious fires occur in existing buildings during maintenance and construction work. Due to the increased fire risks during these periods of time, additional fire precautions may be needed.

Dependent upon the nature of the work to be carried out and the size and use of the workplace, it may be necessary to carry out a new fire risk assessment to include all the new hazards that will be created during the construction work. In some cases the increased risk will be due to the increase of sources of ignition or additional materials. Normally well controlled escape routes may become cluttered by materials.

Attention should be paid to:

- Accumulation of flammable waste and building materials.
- The obstruction or loss of exits and exit routes.
- Introduction of flammable products, e.g., adhesives or flammable gases.
- Materials stored in unusual locations, e.g., the roof, basement or duct areas.
- Boxes in corridors.
- Off cuts of wood and sawdust left in the area where work was done.
- Packing from materials used in the construction process, such as fittings for toilets or cupboards.
- Pallets and plastic covering left near to where materials were used.
- Flammable liquids not controlled - too much stored or stored in unsuitable containers.
- Part empty pots or tubes of adhesives.
- Bulk storage should be outside away from other buildings and sources of heat, including sunlight.

The storage requirements are dependent on the type of material to be stored and reflect the general workplace principles set out in this element. Different material types should be stored separately, to avoid cross contamination and the potential for a harmful adverse chemical reaction. If combined storage is permitted, it is advisable that different materials be kept separated for easy identification and retrieval. In addition to authorised access of site workers, the storage area should be designed to allow safe access by forklift trucks and for the use of suitable mechanical lifting aids to eliminate the need for manual handling where possible. House keeping should be of a good standard and the area kept clean and tidy with suitable lighting and ventilation. There should be appropriate fire precautions and provision of correct fire extinguishers. Storage area access should be restricted to authorised site workers only. The correct safety signs that warn of the hazards and require precautions (no smoking, naked lights prohibitions) should be displayed. Storage areas should be used solely for that purpose and must not be used for any other activities such as mixing operations or as a rest or smoking area.

Figure 3-10: Materials in escape route.   *Source: Lincsafe.*   Figure 3-11: Misuse of storage - poor segregation.   *Source: ACT.*

WASTE DISPOSAL CONSIDERATIONS

Taking into account the size of the site, waste material must not be allowed to accumulate. This means controlling build up of waste at local points around the site, particularly where they impede access and egress, as well as any major accumulation points. The disposal of building materials must be controlled by consignment certificates and taken to a recognised licensed landfill site. Care must also be taken of special waste such as asbestos.

Any waste materials that are left lying around a construction site (or any premises) are a ready source of fuel for any arsonist. Care should be taken regarding the siting of waste skips etc. to ensure that a safe distance between the waste materials and the building exists. In this way, if the waste materials are ignited they should not spread fire to the building itself.

Figure 3-12: Combustible materials.   *Source: ACT.*   Figure 3-13: Materials inappropriately stored.   *Source: ACT.*

DEMOLITION HAZARDS

Structures may contain flammable materials as a result of previous use. Any residues of materials in storage tanks, storage areas or pipelines need to be identified and evaluated. Storage tanks and vessels may contain flammable vapours or toxic sludge's, especially those forming part of an industrial process. Flammable liquids and vapours may also be encountered in confined spaces (such as tanks) and in pipes.

On demolition sites, after the removal of anything valuable or harmful, it is not uncommon for an amount of combustible materials to be left over and for this to be burnt on site to reduce the cost of removal and disposal of material. The safe working procedure for demolition needs to detail the controls to be used for handling, burning and disposing of flammable materials.

USE OF OXY-FUEL EQUIPMENT

Welding

- Only use competent trained staff.
- Regulators should be of a recognised standard.
- Colour code hoses   - blue   - oxygen
                          - red   - acetylene
                          - orange - propane
- Fit non-return valves at blowpipe/torch inlet on both gas lines.
- Fit flashback arrestors incorporating cut-off valves and flame arrestors fitted to outlet of both gas regulators.
- Use crimped hose connections not jubilee clips.
- Do not let oil or grease contaminate oxygen supply due to explosion hazard.
- Check equipment visually before use, and check new connections with soapy water for leaks.
- Secure cylinders in upright position.
- Keep hose lengths to a minimum.
- Close supply valves at cylinder when not in use.
- Follow a permit to work system.

Figure 3-14: Welding equipment.   *Source: ACT.*

Hot work

Hot work has been responsible for causing many fires during construction and maintenance works. One of the most tragic fires due to hot work was the Dusseldorf Airport Fire in 1996. The fire was started by welding on an open roadway and resulted in damage in excess of £200 million, several hundred injuries and 17 deaths.

It is imperative that good safe working practices are utilised. Combustible materials must be removed from the area or covered up. Thought must be given to the effects of heat on the surrounding structure, and where sparks, flames, hot residue or heat will travel to. It is often necessary to have a fire watcher in attendance to spot any fires that may be started. Fire extinguishers need to be immediately available and operatives must know how to use them. The work area must be checked thoroughly for some time after the completion of work to ensure there are no smouldering fires. Strong consideration should be given to the use of hot work permits.

TEMPORARY ELECTRICAL INSTALLATIONS

There is always a danger with temporary installations that standards appropriate to fixed installations will be compromised and hazards will be introduced (e.g. coiled electrical extension leads over-heating, resulting in a source of ignition). To prevent this additional safety measures need to be introduced.

Cables and leads

Any temporary electrical work with cables and leads has the potential to introduce additional fire hazards into premises. To prevent an increased risk you must ensure that insulation appropriate to the environment is used to give resistance to abrasion, chemicals, heat and impact. The insulation must be in good condition as any exposed wire may cause electrical shorting which may cause ignition. Cables must be secured by the outer sheath at their point of entry into the apparatus, including plugs. Again this is to prevent the exposure of bare wires with the potential for sparking. A visual inspection of electrical cables and leads should be carried out before use.

Attention to cables in offices is particularly important to avoid tripping hazards with phone, computer, calculator and kettle leads growing in number. Any tripping over wires firstly poses a health and safety hazard, but secondly may cause physical damage to the cables or cable joints with the potential to cause electrical arcing and/or heat.

Regular examination should be made for deterioration, cuts, kinks or bend damage (particularly near to the point of entry into apparatus), exposed conductors, overheat or burn damage, trapping damage, insulation brittling or corrosion.

3.2 - Arson

Arson as a cause of fire

In 2004, fires due to arson constituted 43% of all fires. The Home Office figures on Arson Control (2004) gave the following facts on arson, for **each week** there are:

- 2,000 deliberately started fires.
- 50 injuries.
- 2 deaths.
- Cost to society of at least £55 million.

Commercial buildings alone have:

- 16,100 fires per year, averaging nearly 310 per week.
- 46 businesses per day are affected by arson.

As part of the management of fire safety within an organisation arson prevention should be a consideration.

Factors influencing the severity and frequency of an arson attack

LOCATION

Many factors affect the potential for an arson attack, one of these being the location of the premises. It would appear that many arson attacks are related to general societal problems. Premises built in a run down inner city area are statistically at much greater risk of arson than if they are sited in a rural setting. The position of the building in relationship to the public roadway and the frequency of people passing by it, will all affect the risk of an arson attack. It is vital that the responsible person for a building or premises keeps abreast of what is happening in the local area. For example, if the adjoining land has becomes wasteland and is then regularly used by the local youths as a playground, this can increase the risk of arson. History has shown that in a short period of time, often resulting from boredom, someone will trespass onto the land and may set fire to rubbish bins, skips, vehicles or the building itself.

SECURITY

Simple but effective ways to deter the arsonist are by giving attention to security, both external and internal, which should encompass the following:

External security
- Control of people having access to the building/site.
- Use of patrol guards.
- Lighting the premises at night.
- Safety of keys.
- Structural protection.
- Siting of rubbish bins at least 8m from buildings.

Internal security
- Good housekeeping.
- Inspections.
- Clear access routes.
- Visitor supervision.
- Control of sub-contractors.
- Audits.

Figure 3-15: Control arson by external security.   *Source: ACT.*

Figure 3-16: Access control, internal security.   *Source: ACT.*

See also - 'Actions to prevent arson attack' as some security measures are also covered within this heading - later in this element.

ACCESS CONTROL

In addition to the general site security access control should be considered as a precaution to limit the risk of arson. This is especially important during the working day when buildings are often unlocked, enabling an arsonist easy access onto the site. Whatever form of access control is installed must however not detract from the requirement for means of escape for anyone inside the building. If the access control equipment (door locks) is worked electronically then it must 'Failsafe' (to open) in the event of a fire alarm being operated, or on power failure. Often this particular point is not physically checked, and there is a risk of people being locked in buildings in the event of a fire.

Actions to prevent arson attack

- Train management to be prepared for arson attacks.
- Address perimeter security with suitable fencing and gates if possible.
- Secure access into the premises - consider roller shutters to external doors/windows.
- Consider active measures such as close circuit television, intruder alarms, night time illumination.
- Never store combustible materials such as pallets, vehicles, against external walls of the premises. Keep external storage to a minimum.
- Ensure skips, etc. are stored as far away from premises as possible - preferably more than 8 metres. Arrange speedy removal of skips.
- Secure wheeled waste bins in designated positions with chains or similar - arrange for bins to be emptied regularly.
- Prevent bushes, tree growth, etc. close to your premises - a dry period will turn such vegetation into a good source of fuel for the arsonist.
- Beware of children gaining access to the site, particularly during school holidays.
- Consider also if your premises front a street:
  - Gaps under doors into a street should be as narrow as possible to stop lighted paper being pushed under them.
  - Install fire resisting mail cabinets to the inside of letterboxes.
  - Air vents and other openings are very vulnerable to attack with flammable liquids particularly if at easily accessible heights. Any such vents should be removed to inaccessible areas or positions.
  - Security should not impede escape for persons in the premises in the event of fire.
  - Always check with your insurance company as they may have specific requirements that must be addressed.
  - Advice should be sought from the Local Authority (Planning Department) in respect of fencing suitability and before fitting roller shutters to windows or doors.

Figure 3-17: Poor siting of skip. *Source: FST.*

3.3 - Prevention of fire

Flammable and combustible materials

General points to prevent fires involving flammable materials in **_use_** within the workplace:

- Quantities of flammable materials in use should be minimal and limited to that which can normally be used by the process in no more than half a day. Any excess quantities must be in a correct storage facility.
- If large quantities of flammables are used in the workplace, piped systems should be considered rather than workers physically handling the products.
- Container lids should always be replaced after use.
- Any rags that have been impregnated with a flammable product will need to be disposed of safely.
- Establish a common earth potential when dispensing / charging containers with flammable liquids.
- Any persons using flammable material should be trained.

General points to prevent fires involving flammable and combustible materials in **_storage_** within the workplace:

- Quantities of flammable materials in storage should be minimal, with sufficient stocks to efficiently run the business, but without overstocking.
- Under traditional guidance you can store up to 50 litres of highly flammable liquids in the workplace in a fire resisting container.
- External storage is preferred to internal storage. If a fire wall is used it should normally provide a fire resistance for a minimum of 30 minutes.
- Storage area should be well ventilated.
- Where possible, containers of flammable liquids should be stored in a bund, drip tray or provided with other spill containment.
- Empty highly flammable liquids and liquefied petroleum gas containers should be stored separately from full ones.
- Prevent container from external damage, for example, impact damage from fork lift trucks.
- Isolate damaged or leaking containers.
- Reduce fire and explosion risk by creating separation between storage containers and limiting the height of stacks.
- Lighting for storage areas for flammable liquids must be 'flameproof' or similar.
- The storage facility needs to be secured from unauthorised access.
- Hazard warning signs should be displayed.
- Any persons storing flammable material should be trained.

General points to prevent fires involving flammable materials in **_transport_** within the workplace:

- If large quantities of flammables are to be transported in the workplace a piped system should be considered.
- Purpose designed metal flameproof containers should be used to transport small quantities of flammable liquids, with lids and anti-spill features.
- Check container lids are secure before moving.
- Palletised containers should be made secure.
- Hazard warning signs should be displayed on containers.
- Establish a common earth potential when dispensing / charging containers for transport of flammable liquids.
- Any persons transporting flammable materials should be trained.

If the material is being transported via a vehicle then transport regulations may apply, making requirements for marking of vehicles, data information sheets available, fire-fighting equipment available and trained staff.

SAFE STORAGE OF FLAMMABLE AND COMBUSTIBLE MATERIALS

When considering the storage of flammable materials the following safety principles should always be applied:

V   **_Ventilation -_** provide plenty of fresh air. Provide ventilation where containers are stored. Good ventilation means vapours given off from a spill, leak, or release, will be rapidly dispersed rather than accumulate to unsafe levels.

I   **_Ignition -_** control of ignition sources. Ignition sources should be removed from the storage area. Ignition sources can vary widely. They include sparks from electrical equipment or welding and cutting tools, hot surfaces, smoking, and open flames from heating equipment.

C   **_Containment -_** use suitable containers and provide spillage control.

Spillages must be contained and prevented from spreading to other parts of the storage area or site. A means of controlling spillage would be the use of an impervious sill or low bund. An alternative is to drain the area to a safe place, such as a remote sump or a separator.

E   **_Exchange -_** consider whether a safer alternative can be used to do the task.

S   **_Separation -_** store away from process areas. Provide separation by a physical barrier, wall or partition where possible.

The mnemonic VICES is a useful reminder of the main principles to be adopted for flammable materials but it should be noted that the most pertinent principle that should be considered first is '***Exchange*** - consider whether a safer alternative can be used to do the task'.

DESIGN AND INSTALLATION OF STORAGE FACILITIES

The design and installation of storage facilities will be dictated by the type and volume of products being stored, plus the desired type of storage, e.g. external or internal. Important factors will be items such as:

- Single storey, light construction in non-combustible materials.
- Space separation of storage from buildings.
- Fire separation between storage and adjoining buildings.
- Sufficient ventilation of enclosed stores by airbricks, vents, forced ventilation.
- Electrical equipment designed for use in flammable atmospheres.
- Consideration of explosion relief.
- Space storage of stocks from each other within storage compound.
- Ullage (air gap above stored liquids) control.
- Door sill and perimeter bunding for containment of up to 110% of the maximum storage container of the chemical.
- Interceptor pits for spillage.

INSPECTION AND MAINTENANCE PROGRAMMES

Housekeeping

Figure 3-18: Examples of poor housekeeping (i).   *Source: ACT.*   Figure 3-19: Examples of poor housekeeping (ii).   *Source: FST.*

By "housekeeping" we mean the general tidiness and order of the building - "a place for everything and everything in its place". At first sight, this may seem a strange matter to discuss when considering fire safety, but as housekeeping affects so many different aspects of this subject, it cannot be ignored. To illustrate how fire safety is affected by housekeeping we must first consider the simplest of fire precautions, that of prevention of fire. Fires need fuel. A build up of redundant combustible materials, rubbish and stacks of waste materials provide that fuel - whether for the arsonists, the carelessly disposed cigarette end / match or the spark from a contractor's welding equipment.

It is not normally possible to eliminate all combustible material from an active workplace, but it should be possible to introduce suitable controls to reduce the risk to an acceptable level. Planned containment and removal of excess rubbish and waste is a major consideration in the risk assessment process. What may be less obvious are the effects of tidiness on fire spread. If a fire starts in a neatly stacked pile of timber pallets, around which there is a clear space, the fire may be spotted and extinguished before it can spread. However, if the same pile were strewn around in an untidy heap, along with adjacent rubbish, the likelihood is that fire would spread over a larger area. In this way it could involve other combustible materials and be very difficult to bring under control without the need for the fire and rescue authority's attendance.

Poor housekeeping does not only affect the ease with which fire can occur, develop and spread, but can lead to:

- Blocked fire exits.
- Obstructed escape routes.
- Difficulty in accessing fire alarm call points/extinguishers/hose reels.
- Obstruction of vital signs and notices.
- A reduction in the effectiveness of automatic fire detectors and sprinklers.

All of the points raised above should be covered during the daily checks, inspections and fire safety audits that should be carried out within the company.

Daily checks

One method of fire prevention which should ensure fire safety is to carry out a simple check of the premises at the *start of the day* which ensures the building is safe to occupy.

This may include points such as:

- Flammable liquids are in suitable containers.
- Electrical equipment is not overloaded.
- Materials are stored away from hot surfaces.
- No deposits or flammable materials on electric motors.
- Waste bins are in place and not overfilled/overflowing.
- Doors used as means of escape are unlocked.
- Escape routes are clear.
- Fire equipment and call points are unobstructed.

This list is not exhaustive; a list relevant to each workplace should be devised.

A *close down check* by an appointed person for each area should be made at the end of each day to ensure the building is safe to leave, and that all necessary security measures have been taken to prevent unlawful entry. This may include points such as:

- Flammable liquids have lids on containers.
- Unnecessary plant and electrical equipment is shut down.
- Materials are stored away from hot surfaces.
- No incipient smouldering fires.
- Waste bins emptied, no accumulation of combustible process waste, packaging materials or dust deposits.
- Waste materials containing flammable liquids have been removed to fire proof bins.
- Fire doors and windows are closed.
- Entry to the building is secure.

Fire safety inspections

Article 17 of the RRFSO establishes a duty to maintain fire safety within the workplace. To assist in maintaining fire safety, regular inspections have been introduced in many companies.

Fire marshals are often appointed from staff and amongst their fire duties they should carry out a fire inspection. They should record any findings, positive or negative, in writing - preferably on a form designed for that purpose. Many organisations devise checklists to prompt the fire marshals to consider an appropriate range of fire issues relevant to the work area under consideration.

By completing the inspection form, and any summary sheets, fire corrective actions and commendations will be highlighted. Such documentation, together with any corrective action carried out, will create an auditable system which will demonstrate management control of fire safety. A copy of inspection forms should be forwarded to the Fire Safety Manager within a formal agreed timescale to enable the Fires Safety Manager to monitor performance and progress on outstanding actions.

A Fire Marshall, or other appointed person, for each area should make an inspection at intervals depending on the fire risks in the workplace - weekly, monthly, quarterly. The inspection will seek to include issues such as:

- Goods neatly stored so as not to impede fire fighting.
- Clear spaces around stacks of stored materials.
- Gangways kept unobstructed.
- No non-essential storage in work areas.
- Materials a safe distance from light fittings.
- 'Hot work' being appropriately controlled.
- Company smoking rules known and enforced.

The process of undertaking fire safety inspections should:-

- Help prevent fires in the workplace.
- Ensure escape routes are clear of obstruction.
- Monitor fire safety standards.
- Keep staff aware of fire safety issues.
- Reinforce the role of fire marshals.

An inspection checklist and report form is a useful control technique to ensure fire safety conditions are identified through inspection and that any actions necessary are collated and rectified. Records should be retained for an agreed period to enable an audit trail to be maintained. An example of an inspection checklist, defect report form and quarterly summary record are provided in figures 3-20 to 3-22.

Fire safety inspection checklist

Weekly / monthly / quarterly

		YES	NO	Defect Rectified
	Escape Routes			
1.	Are exit routes and gangways clear?			
2.	Do exit doors open easily?			
3.	Are exit signs visible and legible?			
4.	Are the fire resisting doors undamaged?			
5.	Do the fire resisting doors shut properly?			
	Fire Fighting Equipment			
6.	Are the fire extinguishers in their correct places?			
7.	Are the fire extinguishers unobstructed?			
8.	Are the fire extinguishers unused?			
9.	Are the fire extinguishers undamaged?			
	Fire Alarm System			
10.	Are the fire alarm call points visible?			
11.	Are the fire alarm call points accessible?			
	Equipment			
12.	Are air vents unblocked on electrical equipment?			
13.	Is electrical equipment being used safely?			
	Staff			
14.	Have all new staff been trained in fire safety?			
	General			
15.	Are the fire hazards unaltered in your area since the last check?			
If any tick has been placed in a clear box above, a report may need to be made.				

Figure 3-20: Fire safety inspection checklist (weekly/monthly/quarterly).

Source: FST.

Fire Safety Inspection defect report

Defect Report

Fire Marshal Area			
Fire Marshal		Date	

Defect Found	

Is Defect Rectified	YES / NO	Is this a repeat problem	YES / NO
Report No			

Figure 3-21: Fire safety inspection (defect report).

Source: FST.

Fire Safety Inspection
Quarterly Summary Record

Date	Correct	Defect Found	Report No	Date Defect Rectified	Comment
Fire Marshall Name			Floor Marshal Area		

Figure 3-22: Fire safety inspection (quarterly summary record).

Source FST.

Fire safety manager - fire safety audit

A systematic review of fire prevention arrangements should be carried out by audit at an agreed frequency, for example once a year, or following a significant event. This would typically be carried out by the Fire Safety Manager. The audit will verify the quality / appropriateness of the fire marshal inspection system and maintenance of general fire arrangements. A fire safety audit will consider the risks, controls and understanding of the arrangements by personnel - in particular the clarity of understanding of roles of those with specific responsibility. Inspections and audit will ensure that fire safety issues within the company are adequately controlled and monitored. Safety assurance should increase as a result, through the implementation of an auditable management system for fire safety.

SAFE WASTE DISPOSAL METHODS

Care must be taken when disposing of items to ensure that fire or explosion hazards are not created. Before disposing of items consider:

■ Does the material or item give off flammable vapours from residues and if so is there a potential for the vapours to be ignited?

■ Is the waste product in a dust format and if this is disturbed can it create a flammable dust cloud?

■ How easy will it be for arsonists to gain access to the waste materials and ignite them, and what effects would this have?

■ Does the equipment/material/product need to be removed by a specialist company?

■ Are additional hazards created by current disposal methods?

Minimising fire risks

SAFE SYSTEMS OF WORK

'Safe System of Work' - an agreement between line management and staff defining how to perform a task safely. This is normally a written procedure, but can be oral dependent upon the level of risk involved.

When designing a safe system of work you need to consider various factors so that fire hazards are not created or the risk of a fire happening increased.

People   The behavioural traits, knowledge, skill, fire risk awareness, level of fire safety training and level of supervision needed and given will all change the effects that people will have on the system that is chosen.

Equipment	Any equipment that is used must be safe to use in the environment concerned and it should be maintained so that new hazards are not introduced, e.g., intrinsically safe equipment in a flammable atmosphere.
Materials	The type of material, or more importantly the manner in which the material is presented in the premises is paramount to fire safety and fire prevention. It is only by understanding the fuel for a fire and its potential ignition, that a safe system of work can be designed.
Environment	Environmental considerations under health and safety requirements would include things such as heating, lighting and ventilation. All these things have the potential to either cause a fire / explosion or to cause unsafe conditions to build up. Again these points must be considered when designing the safe systems.

SAFE-OPERATING PROCEDURES

Procedures should be established to ensure unnecessary fire hazards or risks are not introduced into the workplace without due consideration or control. Consider the introduction of a hot work process into a workplace, perhaps during a maintenance task involving welding. This task would present a high risk if not controlled. As such the task would normally be subject to a hazardous work procedure which requires a formal documented system to be used to control the risk. Such a system is commonly referred to as a 'hot work permit'. ***See also - Permits to work - below.***

PLANNED PREVENTIVE MAINTENANCE PROGRAMMES

As previously discussed, fires can occur due to physical, mechanical and electrical defects in equipment. For example, continuous vibration of a machine can cause electrical connections to work loose, which in turn could create a spark and may start a fire. It is good fire prevention practice to identify such equipment and to schedule planned preventative maintenance at a frequency which prevents such a breakdown. To ensure that this type of maintenance takes place critical equipment should be identified and be included within a maintenance schedule.

MANAGEMENT OF CONTRACTORS

History has proven that having contractors on site can bring an increased risk of fire. Article 20 of the RRFSO makes a requirement on the responsible person to consult with the employer of contractors (or the self employed if relevant) to ensure that they understand the risks and preventive and protective measures that are in place. This process works both ways, so that neither party will cause fire safety issues for the other. It would be as part of this process that the need for a permit to work system would be discussed. Such arrangements need to take account of the use of any sub contractor(s). Careful choice of competent contractors, including their arrangements for fire safety, will be significant issue at the procurement stage.

Many organisations require contractors to carry "Safety Passports", issued through recognised schemes to demonstrate they have received training in issues such as fire prevention. A very thorough scheme is the Client Contractor National Safety Group (CCNSG), a nationally accredited safety passport scheme controlled by the Engineering Construction Industry Training Board (ECITB). This safety passport scheme was developed by major industrial groups and the possession of a safety passport is mandatory for contract workers carrying out work on many sites. The CCNSG scheme covers a range of health, safety, fire and environmental issues. There is a formal assessment, a licence is issued and licence holders details are held on a national register. Passports are valid for three years and refresher training and assessment is required to renew the passport. Sites which operate the scheme have reported a 50% reduction in incidents related to contractors.

PERMITS TO WORK

Out of the range of permits that can be issued in general, the one that affects fire safety issues the most would be a 'hot work permit'. It is recommended that a hot work permit system is used at any point where hot work methods (for example use of a naked flame or creation of hot sparks from grinding) are being used, but not in a place that is designed for such work. A hot work permit would not be required when using a blowtorch in a designated welding booth, because the welding booth should be arranged so that it is safe to do that style of work at all times. However, if a blowtorch is to be used to repair a pipe in a loft space then a hot work permit system would be appropriate to prevent a fire occurring in the roof space.

A hot work permit will ensure that the person carrying out the work does so safely. The permit may include measures such as:

- The removal or covering of all combustibles.
- Other hazards within the area have been identified and controlled.
- Fire-fighting equipment is to hand.
- Knowledge of fire alarm location and fire routine system.
- If necessary an observer to monitor the situation during work, e.g. someone may need to monitor the other side of a wall.
- An authorised person checks and issues a permit prior to commencement of work.
- A competent person signs to accept the permit and agrees to follow the rules imposed.
- A check is made of the area a minimum of 30 minutes after completion of work.
- A competent person returns the permit to the authorised person for signing off.

An example of a typical hot work permit is shown below:

HOT WORK PERMIT
APPLIES ONLY TO AREA SPECIFIED BELOW

Part 1

Site:………………………………………………...... Floor:………………………………………………..

Nature of the job (including exact location) ……………………………………………………………..

……..

The above location has been examined and the precautions listed on the reverse side have been taken.

Date:………………………………………………

Time of issue:…………………………………….... Time of expiry:…………………………………….....

NB. This permit is only valid on the day of issue.

Signature of person issuing permit: …………………………………………………………………….....

Part 2

Signature of person receiving permit:………………………………………………………………………

Time work started:……………………………………………….

Time work finished and cleared up:……………………………………………………………………….

Part 3   **FINAL CHECK UP**

Work areas and all adjacent areas to which sparks and heat might spread (such as floors above and below and on opposite side of walls) were inspected one hour after the work finished and were found fire safe.

Signature of person carrying out final check: ……………………………………………………….....

After signing return permit to person who issued it.

Figure 3-23: Hot work permit - front of form.   *Source: Lincsafe.*

HOT WORK PERMIT
PRECAUTIONS

Hot Work Area

☐	Loose combustible material cleared
☐	Non moveable combustible material covered
☐	Suitable extinguishers to hand
☐	Gas cylinders fitted with a regulator and flashback arrestor
☐	Other personnel who may be affected by the work removed from the area

Work on walls, ceilings or partitions

☐	Opposite side checked and combustibles moved away

Welding, cutting or grinding work

☐	Work area screened to contain sparks

Bitumen boilers, lead heaters etc.

☐	Gas cylinders at least 3m from burner
☐	If sited on roof, heat insulating base provided

Figure 3-24: Hot work permit - reverse of form.   *Source: Lincsafe.*

Maintaining fire protection systems during maintenance and construction work on an existing building

When carrying out works on an existing building there is always the potential for the fire standards and fire protective systems to be compromised. It may be necessary to review the fire risk assessment prior to the commencement of work so that all hazards created by the work can be highlighted and planned for. Common problems encountered would include:

- Accumulation of flammable waste and building materials.
- The obstruction or loss of an exit or exit routes.
- Fire doors propped, wedged open, or missing.
- Openings being created in fire resistant structures.
- Introduction of additional electrical equipment or other sources of ignition.
- Introduction of flammable materials e.g. adhesives.
- Possibilities of false alarms by dust setting off detectors.
- Covers being left on detectors at close of work.
- Disconnection of fire protective system e.g. sprinkler system switched off.
- Introduction of contractors into site who may not be aware of hazards present.

If parts of the fire protection system need to be de-activated during maintenance, it is essential that satisfactory counteracting measures are taken in order to allow the building to remain occupied and to ensure fire safety. For example, if the smoke detection system of a building is taken out of use the counteracting measures may include providing full time active fire wardens to check the building, in particular unoccupied areas. If the fire alarm is de-activated fire marshals may need to be stationed at points throughout the building, with means to raise the alarm and ways to communicate with each other. If counteracting measures are not sufficient to establish a suitable level of fire safety for the building to remain occupied the work will need to be conducted when the building is unoccupied. Measures may be possible sufficient to ensure the fire safety of a small team of workers, even if they are not possible for full occupation.

Fire protection in buildings

Overall Aims

On completion of this Element, candidates will understand:

- effective building design and construction to minimise the spread of fire.
- requirements for adequate means of escape.
- access requirements to assist in fire fighting operations.
- fire detection and alarm systems.
- early fire suppression systems.

Content

Specific Intended Learning Outcomes

The intended learning outcomes of this Element are that candidates will be able to:

4.1   advise on the means of fire protection and prevention of fire spread within buildings

4.2   advise on means of escape

4.3   explain the requirements for ensuring access for the fire service is provided and maintained

4.4   outline the methods and systems available to give early warning in case of fire, both for life safety and property protection

4.5   determine the appropriate level and type of detection within premises, based upon the life and process risk therein

4.6   advise on the selection of basic fire extinguishing methods both for life risk and process risk

Relevant Statutory Provisions

The Regulatory Reform (Fire Safety) Order (RRFSO) 2005

The Building Regulations 2000 Fire Safety - Approved Document B 2000 edition

The Building Regulations 2000 Access to and use of buildings - Approved Document M 2004 edition

Building (Scotland) Act 2003

Health and Safety (Safety Signs and Signals) Regulations (SSSR) 1996

4.1 - Building construction and design - preventative measures

Role of the Building Regulations 2000

The powers to make building regulations are contained in the Building Act 1984. The Act provides that they may be made for the purpose of :

- Securing the health, safety, welfare and convenience of persons in or about buildings.
- For furthering the conservation of fuel and power.
- For preventing waste, misuse or contamination of water.

To this end the current Building Regulations 2000 contain a broad range of what are termed functional requirements with which building work must comply. These requirements cover subjects such as structure, fire safety, sound insulation, ventilation, conservation of fuel and power, and facilities and access for disabled people. They are grouped under thirteen parts (A, B, C, etc) within the Building Regulations.

Definition of elements of structure

The Building Regulations 'Approved Document B' defines elements of structure as:

1. A member forming part of the structural frame of a building or any other beam or column.

2. A load bearing wall or load bearing part of a wall.

3. A floor.

4. A gallery (but not a loading gallery, fly gallery, stage grid, lighting bridge, or any gallery provided for similar purposes or for maintenance and repair).

5. An external wall.

6. A compartment wall (including a wall common to two or more buildings).

However, see the guidance to B3, paragraph 8.4 of the Building Regulations 'Approved Document B' for exclusions from the provisions for elements of structure.

Requirements for fire resistance for elements of structure

The fundamental principle of fire resistant structures is to maintain the integrity of fire compartments in a building and by doing so to prevent the spread of fire. It is important that when a vertical or horizontal compartmenting element is penetrated by a service, the fire resistance of the construction is not decreased.

Fire resistant structures can be found in many forms. A traditional fire resisting structure would be made of brickwork, block work and plaster finishes, which can give in excess of one hours fire resistance. A hollow partition stud wall with a plasterboard finish on either side with a skim of plaster will give 30 minutes fire resistance. More modern building boards can again give 30 minutes plus fire resistance. It is difficult to ascertain if a material is fire resistant just by looking at it and it is necessary, in most cases, to use approved building contractors who provide test certification.

One of the most common failings in building fire resistance is breaches (unauthorised openings) in the fire resistant structure. Breaches will allow fire and smoke to travel unchecked throughout the building, and can often jeopardise the safety of the occupants. It is paramount that whenever any work is undertaken that necessitates holes being made in the fire resistant structure, the structure is made good with fire resistant materials of the same standard as the original structure. This practice is now normally achieved by using intumescent materials. Intumescent materials often consist of surface coatings that are designed to expand when subject to heat, creating foam-like bubbles which act as insulation and therefore reduce the rate of heat transfer.

It is easy to forget the parts of a building not regularly seen by the occupants when considering the need for fire resistance, such as ceiling voids, lift shafts, ductwork, air conditioning systems. These areas can become a major route for fire propagation if they are not installed and maintained such that fire resistance integrity is not compromised.

A problem that has been much focused on in recent times is the potential for fire spread via materials in ceiling and floor voids. This problem has been exacerbated by the increased use of voids to carry electrical, telephone and computer cabling. This presents new fire-paths and fire-load problems associated with the increasing frequency of cabling work as computer systems and telecommunication networks are upgraded and replaced more often to meet the increasing demands of technological developments within organisations.

FIRE RESISTANCE - ELEMENTS OF STRUCTURE

Factors having an influence on fire resistance, which are considered in the Building Regulations, include:

- Fire severity.
- Building height.
- Building occupancy.

It is only by considering these factors that the degree of fire protection that a building needs to have can be assessed. As an example: If the building is multi-storey then it will take longer for people to escape from the building in the event of a fire than if it were a low rise building. As a result the fire resistance may need to be increased so that people are protected from fire for the duration of their escape.

Fire resistant criteria are set for resistance to collapse, fire penetration and transfer of excessive heat.

RESISTANCE TO COLLAPSE

Resistance to collapse refers to the load bearing capacity of an element of a structure and only applies to load bearing elements of the structure. This requirement is often referred to as 'Stability' and in simple words is the ability to withstand a load which is bearing on the element of structure under fire conditions, without failure or collapse of the structure.

FIRE PENETRATION

The requirement for resistance to fire penetration is often referred to as 'Integrity'. This means, in simple words, the element of a structure's ability to prevent a local failure such that under fire conditions fire, and if required smoke, will not penetrate through the structure.

TRANSFER OF EXCESSIVE HEAT

The requirement for resistance to the transfer of excessive heat is often referred to as 'Insulation'. This means, in simple words, the ability to prevent heat transfer through a structure under fire conditions such that the heat will not ignite items on the remote side of the structure.

Compartmentation to inhibit spread of fire within buildings

A fire compartment is a building or part of a building which is designed to prevent the spread of fire to or from another fire compartment. Building Regulations limit the size of compartments in certain types of buildings. However, some buildings, such as single storey factories, have no limitation on the size of compartments. Single storey factory units may therefore consist of one very large compartment, which is why the Fire and Rescue Service often has difficulties in fighting fires in such premises.

Compartmentation is achieved by use of compartment walls and floors which subdivide the building into smaller portions. The majority of buildings utilise traditional methods for controlling fire spread. This is mainly done by the use of fire-resisting structures and fire-resisting doors to break the building into small fire compartments. A fire compartment should withstand a fire for a minimum of 30 minutes, but it may require additional protection, depending upon the purpose of the structure and the use of the building.

If a fire does occur within a compartment it should be confined to that compartment by the nature of the fire resistant materials. This should have the effect of limiting the damage done to buildings and prevent unchecked fire spread. Stairways, ducts etc. should also form separate fire compartments to prevent vertical fire spread. In large compartments sprinkler systems may be fitted in an attempt to limit the size of a fire and ventilation may be provided to allow heat/smoke to escape.

PROTECTION OF OPENINGS IN COMPARTMENTATION

Any opening in a fire compartment structure has the potential to allow fire and smoke to bypass the protection offered by the fire resistance. For example, a set of doors set within a fire compartment structure have the ability to provide a degree of fire resistance. However, this will only be possible if the doors are designed to give the standard of fire resistance required, they are in good condition and closed at the time of the fire. The weaknesses in fire doors as protection is usually people, as we wedge open doors, damage them or do not maintain them correctly. Other protective devices may be items such as steel fire shutters in wall openings and fire dampers in ducting. Fire shutters are often installed into buildings as either an insurance requirement, or as part of the fire compartmentation requirements for building regulation approval. The fire shutter obviously needs to provide the correct degree of fire resistance, but we need to consider the operation of the shutter, its reliability and the potential effects of the shutter closing on any means of escape routes. Fire spread through ducting is a matter of concern. To illustrate this consider the fire at Heathrow airport in December 1997. On this occasion a fire started in the deep fat fryer at a fast food outlet and spread into the extraction ducting. As a result of this fire 2 terminals were closed for business. To prevent such events happening we need to ensure that fire dampers (shutters) that close inside the ducting and thus prevent the passage of fire and smoke are fitted. These dampers should be installed in line with the building's fire breaks so that the fire area is sealed off from the remainder of the building.

FIRE STOPPING

Due to the increased air pressure caused by the heat of a fire on one side of a structure dividing two compartments, the heat and hot smoke from the compartment in which the fire occurs are driven through openings, even small ones, to the other compartment. In order to limit fire spread it is essential that any gaps or openings that breach fire resistant lines are fire stopped. This was traditionally done by the use of bricks / mortar or by installing fire dampers / shutters in ducting. Fire stopping is now achieved by the use of intumescent materials to fill gaps or intumescent grills in ducting.

Figure 4-1: Opening not fire stopped.   *Source: FST.*   Figure 4-2: Fire stopping carried out correctly.   *Source: FST.*

PROTECTION OF CONCEALED SPACES (CAVITIES)

Another factor to consider in the prevention of fire spread within buildings is the control of large voids within the building. Fire will quickly spread unnoticed through any large concealed space, for example, an open ceiling void above a false ceiling. To prevent fire spread in this way a form of compartmentation of the cavity is carried out. This is often done using fire retardant blankets which are suspended from the upper face of the void and span the entire void space. The material often used for the fire retardant blankets is rock wool mineral fibre. In this way a fire barrier is created which provides a degree of compartmentation of the cavity and limits fire to spread.

Internal fire growth and lining materials

SANDWICH PANELS

In modern buildings one type of construction which is commonly used, but is causing great concern, is 'Sandwich Panels'. These consist of two outer skins of sheet metal (normally a light alloy) with an infill of heat insulating material. In some cases the insulating material used is polyurethane or styrene foam.

This type of construction is causing problems in buildings when they are on fire due to sudden and unannounced building collapse as the panels fall out of their framework. In addition, they can cause very rapid fire spread once the internal foam is on fire as the fire spreads inside the sandwich panels.

As a result of these problems, the Fire and Rescue Services may have to consider attacking a fire in a building of this type from outside the building, in order not to endanger the lives of firefighters.

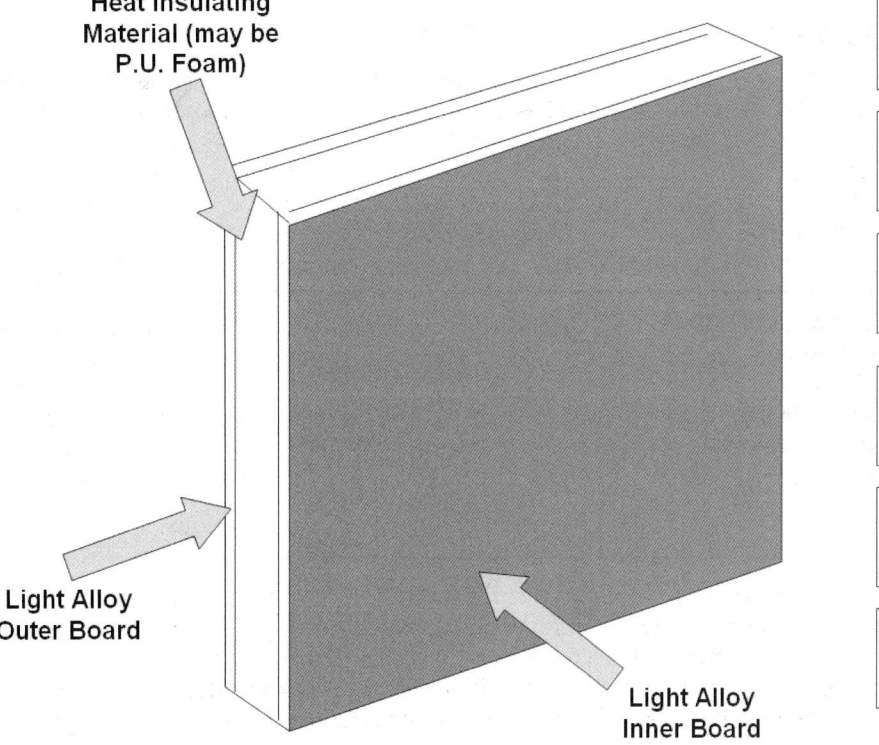

Heat Insulating Material (may be P.U. Foam)

Light Alloy Outer Board

Light Alloy Inner Board

End Supports

Exposed Foam

Holes Cut Through

Workmanship

Structural Collapse in Fire

Fire Brigade Non-attack Strategy

Figure 4-3: Sandwich panel.   *Source: FST.*

BUILDING BOARDS AND SLABS

The characteristics of fire resistance and surface spread of flame are inherent qualities of board materials. If the performance qualities of boards are low it may be possible to increase their qualities by the use of a fire resistant substance which is either attached to or impregnated in the board material. As part of the fire risk assessment process we need to consider both the speed at which fire will spread across the surface of a building material and the effects or contribution if any the material would have on the fire itself. As an example: a building board with metal light alloy facing covers will have an excellent fire spread characteristic as the metal sheet will not burn. However, if the filling inside the building board is polyurethane foam this will decompose with the heat from the fire, give off large quantities of flammable gasses which can then cause a flashover in the building as the fire propagates through the flammable gas layer.

Testing systems for the surface spread of flame characteristic gives an indication of the speed that fire would spread across a material's surface. Class 1 materials would present the slowest speed of flame spread and class 4 the fastest. Class 0 is often mentioned in literature; it is not a true classification, but to be categorised as Class 0 a material must be Class 1 and must not contribute greatly to the propagation of the fire.

Escape routes and circulation spaces within buildings should have both ceilings and walls comprising materials of Class '0' standard. Some examples of materials with different classification are given below:

Plasterboard	Class '0'
Woodwool slabs	Class '0'
Mineral fibre board	Class '0'
Chipboard	Class '3'
Softboard	Class '4'
Plywood	Class '3'

Figure 4-4: Material classifications. *Source: FST.*

WALL LININGS

Room linings should not be easily ignitable. Consideration should be given to the effects of the wall linings on the growth of a fire as it may lead to a flashover occurring.

This is illustrated in the table below, which is extracted from test results achieved by the Building Research Establishment.

Wall Lining	Flashover Time
Dense non-combustible material, e.g. brick	23 minutes 30 seconds
Fibre insulating board with skim of plaster	12 minutes
Hardboard with 2 coats of flat oil paint	8 minutes 15 seconds
Non-combustible insulating material	8 minutes

Figure 4-5: Wall lining test results. *Source: Building Research Establishment.*

As can be seen, even materials that are classed as being non-combustible can affect the development of a fire due to their insulating properties raising the temperature within the compartment.

Means of preventing external fire spread

The construction of external walls and the separation between buildings to prevent external fire spread are closely related. The likelihood of a fire spreading across an open space between buildings depends on:

- Size and intensity of the fire in the building concerned.
- Distance between the buildings.
- Fire protection given by the facing sides of the buildings.

The requirements for the control of external fire spread will be met:

- If the external walls of the building presumed to be on fire are constructed of a material that prevents or reduces the risk of ignition from any external source, and the spread of fire over their surfaces will be limited.
- If the amount of unprotected openings in the side of the building presumed to be on fire is limited so as to reduce the amount of thermal radiation that can pass through the wall and affect an adjacent building. When doing this the distance between the wall, the boundary and adjacent buildings is taken into account.
- If the roof of the adjacent building is constructed so that the risk of spread of flame and/or fire penetration from an external fire source is limited.

The objective of each of the above factors is to limit the risk of a fire spreading from one building to another. The extent to which this is necessary is dependent on the use of the building, its distance from the boundary and, in some cases, its height.

CONSTRUCTION OF EXTERNAL WALLS AND ROOFS

To protect from external fire spread the structure must be able to withstand the effects of fire, as described above. In the majority of cases the minimum fire resistance for the elements of structure is 30 minutes and is often 60 minutes. However, this criteria is linked directly to the distance of the building from adjoining buildings or from the property boundary line.

DISTANCE BETWEEN BUILDINGS

As mentioned above another factor that is considered when assessing external fire spread is the space separation between buildings, or between the building and the property boundary. The critical distance is one metre and if space separation is less than this distance then building regulations will impose various requirements on the nature and type of structure and materials that can be used. Another factor that is dictated by the space separation of buildings is the amount of allowable unprotected openings in the façade of the building. An unprotected opening is any part of the external wall which has a lower fire resistance than the minimum requirement for the wall itself, for example, a window in a brick wall. Again the requirements are stricter if the space separation is less than one metre.

4.2 - Means of escape

Definition of means of escape

The following is a widely accepted definition of means of escape:

> *"Structural means whereby (in the event of a fire) a safe route or routes is provided for*
> *persons to travel from any point in a building to a place of safety."*

Figure 4-6: Means of escape - definition.   *Source: Building Regulations Approved Document B.*

A careful study of each component in this definition will show that it can be used to guide towards the provision of suitable and sufficient means of escape.

STRUCTURAL

Good means of escape must be part of the structure of the building and made immovable. The effect of this is to generally rule out the use of most portable self rescue devices, for example, roll out ladder, which cannot be relied upon to be in position when required.

TRAVEL

The word 'proceed' instead of 'travel' may better indicate what is required. Ideally it should be possible to 'turn one's back on the fire' and walk away to a safe place. This should be possible from all parts of the building.

FIRE

The definition emphasises that in the event of a fire persons should be able to safely escape. It is possibly less clear that this must include the fire itself as well as smoke and toxic gases arising from it. Protection against the products of fire is probably the most important part of means of escape.

PLACE OF SAFETY

A place of safety is ideally in the open air from where dispersal can take place. This would be termed a place of 'ultimate safety'.

In more complex buildings it is not always possible to immediately escape to a place of 'ultimate safety', for example, when evacuating high buildings. In these cases escape will be made initially to a place of 'comparative safety', such as a fire protected stairway. The people would then continue their escape using this fire protected stairway until they were able to reach a place of 'ultimate safety'.

The principles of means of escape and general requirements

In the main, guidance on acceptable means of escape can be found in published British Standards (BS), for example BS 5588 Pt II Fire Precautions in the Design, Construction and Use of Buildings', and Government Department Guides.

Though these documents provide guidance on the provision of means of escape the application of the guidance has been influenced by practical experience over many years. From this experience the following general principles have evolved. The explanation and examples provided are for general illustration purposes only, and have no legal standing. Individual Guidance books, Codes of Practice or British Standards and risk assessment techniques should be used to assess the means of escape necessary in any specific buildings.

The general principles for the design of means of escape are:

■ That there should be alternative means of escape from most situations.

■ Where direct escape to a place of safety is not possible, it should be possible to reach a place of comparative safety, such as a protected stairway, which is on a route to an exit and is within a reasonable travel distance.

In such cases the means of escape will consist of two parts - the first being unprotected in accommodation and circulation areas and the second in protected stairways (and in some circumstances protected corridors). The place of ultimate safety is the open air clear of the effects of the fire. However, in modern buildings which are large and complex, reasonable safety may be reached within the building, provided suitable planning and protection measures are incorporated.

The following are not acceptable as means of escape:

■ Lifts (except for a suitably designed and installed evacuation lift that may be used for the evacuation of disabled people, in a fire).

■ Portable ladders and throw-out ladders.

■ Manipulative apparatus and appliances, for example fold down ladders and chutes.

Escalators should not be counted as providing predictable exit capacity, although it is recognised that they are likely to be used by people who are escaping. Mechanised walkways could be accepted and their capacity assessed on the basis of their use as a walking route while in the static mode.

ALTERNATIVE ESCAPE ROUTES

There is always the possibility of the path of a single escape route being rendered impassable by fire, smoke or fumes. In this situation, ideally, people should be able to turn their backs on a fire wherever it occurs and travel away from it to a final exit or protected escape route leading to a place of safety. However, in certain conditions a single direction of escape (a dead end) can be accepted as providing reasonable safety. These conditions depend on the:

■ Use of the building and its associated fire risk.

■ Size and height of the building.

■ Extent of the dead end.

■ Numbers of persons accommodated within the dead end.

MAXIMUM TRAVEL DISTANCES

Travel distance has been defined as:

> *"The distance to be traversed in order to reach a place of safety"*

Figure 4-7: Simple travel distance definition.   *Source: FST.*

Place of safety in this context can be either ultimate or comparative. If evacuation times are to be maintained, then some limit has to be placed on the travel distance acceptable. It is not possible to set maximum distances which universally apply to a variety of buildings with different occupancies, so it is essential to refer to specific codes of practice for guidance. This would include maximum distances to:

■ Final exit.

■ A staircase which is a protected route.

■ Door to a protected lobby.

■ Door to an external escape route.

Codes of practice set out guidance on reasonable travel distances for a given situation. When coming to a conclusion on travel distances for a specific building using a risk based approach the code of practice should be used as a guide for setting maximum distances. This should only be exceeded if a suitable and sufficient risk assessment has determined that a less than average risk can be established. For example, if a building has a very high ceiling this may be assessed as presenting a reduced risk and therefore a longer than average travel distance may be considered reasonable.

NUMBER AND SIZE OF ESCAPE ROUTE FOR NUMBER OF OCCUPANTS

Entrances, exits and circulation areas are provided in all buildings for normal everyday use and a means of escape should utilise, where possible, existing arrangements. The first approach to provision of means of escape should therefore be to look at the existing exits, their locations, number and width. Only if they are insufficient in some respect should further steps be taken. The minimum width of an exit should be 750 mm, more than one exit should be provided if there are more than 60 people in a room.

When the question of adequacy, in number or width, arises a more detailed study of the movement of persons becomes necessary. This is usually so when dealing with places of public resort or those having a high density factor. In most other cases the width of exits is not a crucial factor as the number provided for normal use is generally adequate to cope with the number of persons involved. The number of escape routes and exits that are required depends on the number of occupants in the room, its level above or below ground and any limits on travel distance to the nearest exit. When travel distances are measured, they are measured to the nearest storey exit only.

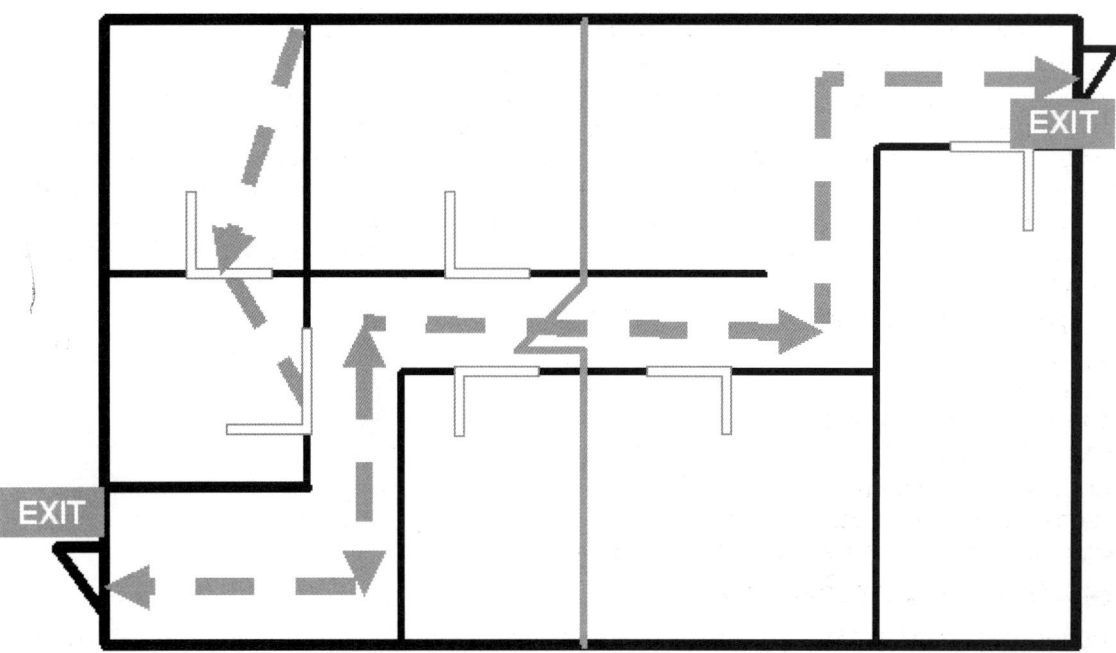

Figure 4-8: Diagram depicting both one way and two way travel. *Source: FST.*

BASIC OCCUPANCY CALCULATIONS

In existing occupied buildings the number of current occupants can be determined by counting the people, however the number may not remain constant and this is not always possible or suitable for new buildings.

To overcome this problem 'floor space factors' have been calculated. These work on the basis that, depending on the type of situation, a person will require a certain amount of space, calculated in square metres of area (m^2). This is calculated empty - before items/storage is put in.

Typical occupancy factors are:
- Dining room, committee room   1.0 m^2
- Offices   6.0 m^2
- Standing area in bars   0.3 m^2
- Art gallery, dormitory, factory production area, museum or workshop   5.0 m^2

REQUIREMENTS FOR ESCAPE STAIRS

An important aspect of means of escape in multi-storey buildings is the availability of a sufficient number of adequately sized and protected escape stairs.

Number of escape stairs

In general there will be more than one escape stairway. However, the number of escape stairs needed in a building, or part of a building, will be determined by:

- Constraints imposed due to the design of horizontal escape routes, e.g. travel distances.
- Whether independent stairs are required due to a mixed occupancy building use.
- Whether a single stair is acceptable.
- Provision of adequate width for escape. The width is calculated either on the basis of a stairs having to be discounted or otherwise. This decision is made based on the level of fire protection given to the stairs and the possibility of a fire on one of the floors spreading smoke into a stairwell.

In larger buildings provisions for access for the fire service may be necessary, in which case some stairs nominated for escape may also need to serve as fire fighting stairs. The number of escape stairs provide may need to be increased to take account of this.

Width of escape stairs

The width of escape stairs should:

- Be not less than the width required for any exits giving access to the stairs.
- Have a minimum width of 1 metre.
- Not be too wide so that people spread out during the evacuation (Building Regulations state that the width of stair should not exceed 1400mm unless there is a central handrail).
- Be fitted with a handrail if wide.
- Not reduce in width at any point on the way to the final exit.

General points

■ Escape stairways within a building should be enclosed within a fire resistant structure.

■ Fire doors should not be wedged open.

■ Stairs should lead direct to fresh air, or to two totally separate routes of escape.

■ Steps and treads to be non slip/trip and to be in good condition.

■ Stairs should incorporate handrails singly if narrow stairs, either side if double width and central handrail if over 1400mm wide.

■ There should be no storage of combustible materials within or under the staircase.

Figure 4-9: Rear escape stairway obstructed.   *Source FST.*

PASSAGEWAYS AND DOORS

Passageways

An integral part to the escape routes in the majority of premises is the corridors, passageways and doors. One of the main considerations should be keeping escape routes clear from obstructions and blockages. Consideration needs to be given to any fire hazard that is located on an escape route passageway, especially if there is only a single escape route available. Both the passageways and corridors need to be wide enough for the number of people who may need to use them. The corridors should lead direct to fresh air, or to a fire protected escape route.

Figure 4-10: Storage on escape route.   *Source: FST.*

Figure 4-11: Unobstructed marked escape corridor. *Source: FST.*

Doors

Fire doors serve two purposes:

■ They prevent the spread of fire and smoke.

■ Ensure means of escape for people using the building.

Escape doors (doors through which a person passes whilst escaping) should:

■ Open in the direction of escape (unless low occupancy numbers and assessed as acceptable).

■ Should not be wedged open.

■ Should be easily and immediately operable ('easy' meaning no resistance to opening).

■ Should lead to a point of safety.

■ As with the exit route, the doors need to be wide enough to evacuate the number of people who may need to use them.

It is often difficult to tell visually what is a fire door. However, points to look for are:

■ The door will normally be fitted with three hinges.

■ The door will normally be fitted with a positive self closing device, e.g. swing arm device or percomatic closer.

■ New fire doors will be fitted with intumescent strips and cold smoke seals.

■ New doors should have a colour rawl plug inserted into one of its side edges. The colours on the plug denote the standard of the door, e.g. white circle with red dot is an FD20/20 door which requires intumescent materials to be fitted. A white circle with a green dot is an FD20/20 door which does not require intumescent materials.

There are a few terms used in the statements above with which some people may not be familiar.

Swing arm device

Traditional type of door closer which fits at top of door and joins onto door frame. Once opened it should slowly close and latch the door.

Percomatic closer

A type of door closer which uses spring loaded pistons for its methods of operation. The door closer itself is mounted inside the leaf of the door after a hole has been routed out. The only visible portion is a single or double chain between the door and door frame (hinge side). Again this device is intended to close and latch the door.

Intumescent strips

Intumescent material is material that expands when it get hot. Strips of this material are fitted around both sides and the top of the door. If there is a fire the material intumesces (swells) and fills the gap between the door and door frame, thus preventing fire and smoke from passing through.

Cold smoke seals

Intumescent materials work well if a fire is close to a door or well developed. If the fire is small or distant from the door, by the time the smoke reaches the door it is cold and the intumescent material will not intumesce. Smoke would then pass around the edge of the door.

A cold smoke seal is a nylon brush or neoprene strip which again is introduced both sides and at the top of the door. This seal should constantly brush against the door/door frame and therefore prevent the cold smoke from passing through the gap.

FD20/20 and FD30/20 etc.

Fire doors are now denoted as shown above. FD stands for fire door. The first figure stands for stability test rating in minutes, the second figure stands for integrity in minutes.

- Stability - is the time at which collapse occurred.
- Integrity - is the time that cracks or other openings exist through which flame or hot gases can pass and cause flaming of a cotton wool pad.

The major point to bring out in the use of fire doors is that **they should not be wedged open.**

When considering fire doors and their importance, it may be worth reminding ourselves of the requirements we have to meet. Below are extracts from the Regulatory Reform (Fire Safety) Order (RRFSO) 2005:

Emergency routes and exits

Where necessary in order to safeguard the safety of relevant persons, the responsible person must ensure that routes to emergency exits from premises and the exits themselves are kept clear at all times. The following requirements must be complied with in respect of premises where necessary (whether due to the features of the premises, the activity carried on there, any hazard present or any other relevant circumstances) in order to safeguard the safety of relevant persons -

- Emergency routes and exits must lead as directly as possible to a place of safety.
- In the event of danger, it must be possible for persons to evacuate the premises as quickly and as safely as possible.
- The number, distribution and dimensions of emergency routes and exits must be adequate having regard to the use, equipment and dimensions of the premises and the maximum number of persons who may be present there at any one time.
- Emergency doors must open in the direction of escape.
- Eliding or revolving doors must not be used for exits specifically intended as emergency exits.
- Emergency doors must not be so locked or fastened that they cannot be easily and immediately opened by any person who may require to use them in an emergency.
- Emergency routes and exits must be indicated by signs.
- Emergency routes and exits requiring illumination must be provided with emergency lighting of adequate intensity in the case of failure of their normal lighting.

PROTECTION OF ESCAPE ROUTES

It is often necessary to fire protect the escape routes so that people evacuating have a more assured way out. In general it is necessary to fire protect stairways and corridors that are the only means of escape from an area. If a corridor gives escape in both directions, it may not be necessary to fire protect it.

However the application of this approach must be based on risk, for example, if the corridor giving escape in both directions is in an hotel it will probably need to be fire protected to give sufficient protection to people trying to escape the building. The decision to fire protect the escape route in hotels is based on the risk of delay in evacuating such buildings due to the tendency of hotel residents to be delayed because they are either asleep or resisting leaving.

EMERGENCY LIGHTING AND SIGNAGE

After the design and provision of means of escape has been concluded, its safe and effective use in case of fire must be established and maintained.

Factors that have to be considered with escape routes are the amount of natural light along the route and the time of day it is occupied. If an escape route has no natural or borrowed light, or it is occupied out of normal day time hours, then escape lighting should be required.

Where use of an escape route is dependent upon artificial lighting, it is usual to install a system of secondary lighting, known as emergency or escape lighting. Provision should be such that it will adequately illuminate escape routes in the event of failure of the normal electrical supply to the lighting circuit which may arise, for example due to fire.

Article 14 of the RRFSO requires "emergency routes and exits requiring illumination must be provided with emergency lighting of adequate intensity in the case of failure of their normal lighting".

Under other legislation the provision of emergency lighting may also be required, either directly as in B1 of the Building Regulations or as a result of conditions imposed by licences for high risk premises.

Emergency lighting is usually provided on common access routes in commercial, industrial and multi storey premises.

Functions of emergency lighting

Emergency lighting is required to fulfil the following functions:

- Indicate clearly the escape routes.
- Provide illumination along such routes to allow safe movement towards and through and ideally beyond the exits provided.
- Ensure that fire alarm call points and fire fighting equipment provided along escape routes can be readily located.

One other consideration may also be the use of photoluminescent signs and markings, which are especially of use if the existing escape lighting system gives a low light level. It may be possible through risk assessment to use photoluminescence to improve vision sufficiently, as an alternative to upgrading an existing system.

Emergency lighting can be subdivided into different types as shown on the diagram below:

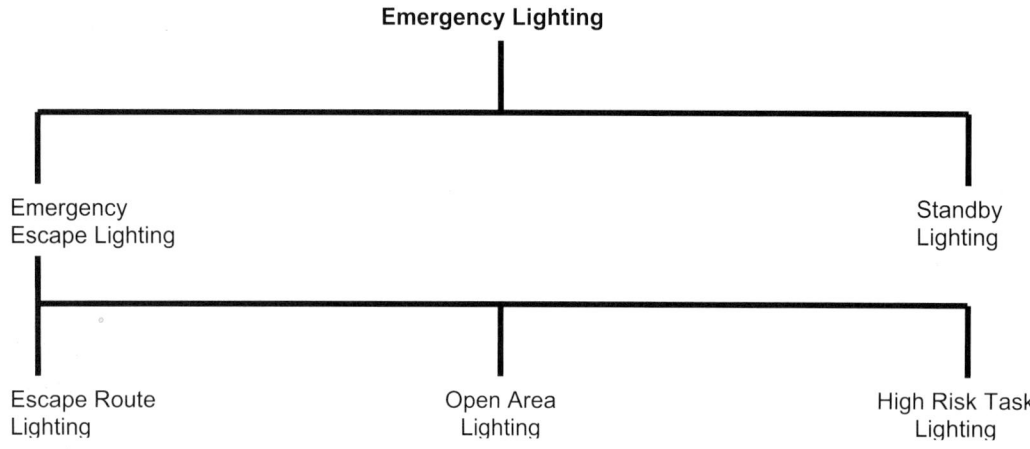

Figure 4-12: Emergency lighting.   *Source: FST.*

Types

Emergency escape lighting	The part of the emergency lighting that provides illumination for the safety of people leaving a location, or attempting to terminate a potentially dangerous process before doing so.
Escape route lighting	The part of the emergency escape lighting that provides sufficient light to ensure the escape route can be effectively identified and safely used.
Open area lighting	The part of the emergency escape lighting provided to avoid panic and provide illumination allowing people to reach a place where an escape route can be identified. Sometimes known as 'Anti Panic Lighting'.
High risk task lighting	The part of the emergency escape lighting that provides illumination for the safety of people involved in a potentially dangerous process or situation and to enable proper shut down procedures for the safety of the operator and other occupants.
Standby lighting	The part of the emergency lighting provided to enable normal activities to continue substantially unchanged.
Maintained	Operates at all times.
Non Maintained	Only lights up when power fails.
Sustained	Contains 2 lamps - 1 powered from mains only, 1 also powered from emergency lighting supply (usually batteries).
Self Contained Luminaire	Luminaire containing everything - lamp, battery, test facility and controls.

Position of lighting luminaires

Luminaires should be found in the following locations:

- Exit doors intended to be used in emergency.
- Near stairs so that each flight is illuminated.
- Near any other change in level.
- Mandatory emergency exit and safety signs.
- Each change of direction.
- Each intersection of corridor.
- Outside and near to final exit doors.
- Near to each fire fighting equipment point.
- Near to call points.
- Near to each first aid point.
- Lift cars.
- Moving stairways and walkways.
- Toilets, lobbies and closets over 8m^2.
- Motor generator, control and plant room.
- Pedestrian escape routes from covered car parks.
- Plus anywhere else as required to reach minimum level of lighting required.

Figure 4-13: Luminaire. *Source ACT.*

Lighting levels

Defined escape route   e.g. corridor, stairs, 0.2 Lux on centre line of escape route, minimum duration 1 hour.

Open area lighting   e.g. open office, 0.5 Lux all over floor area except for 0.5m border, minimum duration 1 hour.

High risk task lighting   e.g. rotating machine, 10% of normal light level (minimum 15 Lux) in max 0.5 seconds, minimum duration - for as long as risk exists.

Testing of emergency lighting system

Daily

1. Check no faults - normal power.
2. If maintained system, check lamp is lit.
3. Any faults to be rectified and action recorded.

Monthly

1. Simulate a power failure to the normal lighting circuit and check that self contained luminaires have energised.
2. Simulate a power failure to the normal lighting circuit to check that central battery system energises (if a central battery system is fitted).
3. Simulate a power failure to lighting to check that generator starts up and energises lights for a minimum of one hour (if generator system fitted).

6-Monthly

Longer duration tests should be done, e.g. 3 hour self contained luminaires should be energised from battery for 1 hour.

There are also additional tests after 3 years and thereafter annually.

SIGNAGE

Introduction

Fire Safety signs have been a point of debate over recent years, with the conflict between what has been acceptable under British Standards over many years and the need for a European standard. The Health and Safety (Safety Signs and Signals) Regulations (SSSR)1996 have tried to clarify this.

There are two acceptable types of fire safety signs:

- EC Directive Signs - the European standard sign is a white pictogram on a green background, it has only a plain white block, which denotes an exit. This may have directional arrows showing the way to an exit.

Figure 4-14: Sign to comply with EC Directives.   *Source: FST.*

- BS 5449 - these signs are a white pictogram on a green background and depict a running person framed in a doorway, which is an exit. These may be seen with or without directional arrows pointing the way to an exit.

Figure 4-15: Sign to BS 5499 Part 4.   *Source: FST.*

Other types of signs which are to be used in premises are descriptive, e.g. the 'mandatory' signs that set a requirement for action 'FIRE DOOR - KEEP SHUT' and 'KEEP LOCKED SHUT'. Fire procedure notices also come into this category and are in white text on a blue background (or may be reversed), and normally carry the following information:

- Method of sounding the alarm.
- Method of calling the Fire Brigade.
- Method of evacuation, e.g. 'LEAVE BY NEAREST EXIT'.
- Location of assembly point(s).
- Specific instructions, e.g. 'DO NOT RE-ENTER BUILDING', 'DO NOT COLLECT PERSONAL BELONGINGS', 'DO NOT USE LIFTS' etc. This portion of sign must be red/white to signify 'prohibition'.

Fire equipment signage

This type of fire signage is used to identify the location of fire equipment. These are used to indicate fire alarm call points, fire extinguishers and hosereels. These are coloured red with a white pictogram depicting whichever piece of equipment is in position.

Figure 4-16: Fire action notice   *Source: FST.*

Figure 4-17: Well signed fire extinguishers.   *Source: ACT.*

Application and use of signs

The objective of the escape route signing system is to ensure that exits are identified from any place within a building. Where direct sight of an exit is not possible and doubt may exist as to its position, a directional sign (or series of signs) is provided. The signs need to be positioned in such a way that people can follow the escape routes to the final exits. If there is a choice of escape routes from a given point, the signage system should denote the route with the shortest distance. Directional arrows and supplementary text may be added to signs as necessary.

In some premises, e.g. care homes, there is a need to make the surroundings of residents appear less institutional and therefore directional signs may be used less frequently. Staffing levels and training would need to be greater in such instances to compensate.

A general rule is that you should be able to see your way to an alternative exit from any point in the building and therefore in large factories exit signs may need to be at high level for greater visibility.

To ensure that signage is adequate the following needs to be considered:

- Sufficient in the building to ensure people can escape.
- Positioned in the correct places.
- Visible.
- Conform to the current standards.

Supplementary text

It is acceptable to add supplementary text to signs to assist with understanding, for example:

- ***Exit -*** denotes a doorway or opening that leads to a place of ultimate safety
- ***Fire exit -*** denotes a doorway or opening that leads to a place of safety, which has been specifically provided as an alternative exit to be used in the event of the evacuation of the workplace. This does not prevent the door being used for everyday use.

Figure 4-18: Poorly signed fire escape.   *Source: FST.*

Size of sign

The size of sign is dependent upon the viewing distance from which the sign is seen and the type of sign, for example whether it is a standard, photoluminescent, internally illuminated or externally illuminated sign. In simple language, it needs to be big enough to be seen and do its job. However, care must be taken with very large signs as this may cause people to move towards the exit they are denoting and by doing so missing nearer exits that have smaller signs.

Directional arrows

Due to recent changes, there is confusion over which direction the signs are denoting. The British Standard version signs are as follows:

Forward from here
Forward and through from here
Forward and up from here

Progress down from here

Progress right

Progress left

Progress up to right, or
Forward and across to right (in open areas)

Progress down to right

Progress up to left, or
Forward and across to left (in open areas)

Progress down to left

Figure 4-19: Directional arrows.   *Source: BS 5449.*

DESIGN FOR PROGRESSIVE HORIZONTAL EVACUATION

The concept of progressive horizontal evacuation allows progressive horizontal escape to be made by evacuating into adjoining compartments, or sub-divisions of compartments. The object is to provide a place of 'comparative' safety within a short distance, from which further evacuation can be made if necessary but under less pressure of time.

When this system of evacuation is adopted in a building the integrity of the fire resistance and compartmentation needs to be assured for sufficient time to conduct the progressive evacuation. This system would be used in hospitals, care homes and other similar premises.

Management actions to maintain means of escape

Management needs to ensure that the means of escape from premises are maintained in an efficient state, in good working order and in good repair. Article 17 of the RRFSO establishes a requirement for this. In addition, Article 14 of the RRFSO sets out a requirement "Where necessary in order to safeguard the safety of relevant persons, the responsible person must ensure that routes to emergency exits from premises and the exits themselves are kept clear at all times". The best way of achieving this is by implementing a good system of fire safety checks and inspections and by profiling a thorough system of planned maintenance and testing for all aspects of the premises that have an effect on the means of escape.

Requirements for means of escape for disabled people

Due to recent changes in the Discrimination Disabled Act (DDA) alterations may need to be made to enable people with disabilities to access buildings. This may mean that people with disabilities may be found on floors where they were not previously able to gain access. The fire management system should ensure that **all persons** who enter buildings can escape in the event of a fire. There is a misconception in some companies that it is satisfactory to evacuate the person with disability to a safe refuge e.g. fire protected stairwell and to hold them there awaiting the arrival of the Fire and Rescue Service, so that they can carry them down - this is not the case. It remains the responsibility of the 'responsible person(s)' for the premises to ensure all who are present can be evacuated, without assistance from the Fire and Rescue Service, should the need arise.

This issue is not however straightforward as there are many forms of disability which may be encountered. The more common disabilities would be mobility impairment, vision impairment and hearing impairment. All the strategies set out below must be supported by active fire marshals that are made aware of the presence of the people with disability.

Hearing impairment	Evacuation may be assisted by the issue of personal trembler alarms, flashing lights and 'buddy' (work companion assistance) systems or the equivalent for visitors to the building.
Vision impairment	Evacuation may be assisted by use of tactile way-finding and exit signs and 'buddy' systems or the equivalent for visitors to the building.
Mobility impairment	This may vary from someone who is just slower than everyone else in escaping to a person in a wheelchair. Evacuation methods may vary from assisting the person out of the building immediately after the initial rush of occupants to use of evacuation chairs and other escape systems. The worse case scenario should be considered such as the need to evacuate a person in a wheelchair vertically, where the risk factors involved in moving them would need to be known and provided for. An interim step might involve a responsible person remaining with the individual(s) in a designated safe refuge (a place of comparative safety), and the responsible person would then communicate their location at this point so that an informed judgement can be made on the need to evacuate the individual further or not.

The actions required to ensure the safe and effective evacuation of disabled people in an emergency situation need to be given detailed consideration. Management procedures need to be in place that take account of the various scenarios that may arise. For example, procedures adopted with regard to disabled people employed in the building may well be different from those for disabled people visiting the building who are therefore unfamiliar with its layout.

BS 5588 'Fire Precautions in the Design, Construction and Use of Buildings' states:

> *"It is neither possible nor desirable ... to recommend which procedure should be adopted in any particular circumstances. Circumstances will vary as to the needs of disabled people and whether their relationship with the building management is a continuing or transient one".*

Figure 4-20: Needs of disabled people.   *Source: BS 5588.*

Systems of evacuation that may be implemented include:

- Horizontal evacuation.
- Evacuation by lift.
- Evacuation by stairs.
- Use of refuges.

USE OF EVACUATION LIFTS AND REFUGES

Lifts

An evacuation lift is a lift which has been specifically designed for the evacuation of disabled persons. In essence these lifts are set within a fire resisting enclosure and have a separate power supply so that their use can be assured during a fire. Evacuation lifts are often utilised in conjunction with 'refuges'.

Refuges

BS 5588: Part 8 defines refuges as:

"An area that is enclosed with fire-resisting construction (other than any part that is an external wall of a building) and served directly by a safe route to a storey exit, evacuation lift or final exit, thus constituting a temporarily safe space for disabled people to await assistance for their evacuation.
NOTE Refuges are relatively safe waiting areas for short periods. They are not areas where disabled people should be left alone indefinitely until rescued by the fire brigade or until the fire is extinguished".

In this situation a refuge is an area that is both separated from the fire by a fire-resisting construction and which has access via a safe route to a fire exit. It provides a temporary space for people to wait for others who will then help them evacuate. Where people reach the relative safety of a refuge prior to vertical movement it will be necessary to ensure that there is adequate space in the refuge so that others evacuating are not obstructed.

The use of refuges should always be viewed within the context that buildings should have adequate measures in place to enable the evacuation of all persons.

Evacuation by stairs

At some point it may be desirable or necessary to move people with disabilities down the evacuation stairs. When this involves going down stairs or steps rather than going up, one of the most common systems used is the 'Evac+chair', Evac+chairs can be used by people trained and capable of using the equipment. It is worth noting however that additional carry handles can be added to Evac+chairs, or other specialist chairs purchased that will allow for evacuation up stairs.

USE OF GRAPHIC, AURAL AND TACTILE WAY-FINDING AND EXIT SIGN SYSTEMS

As part of the control measures put in place for safe evacuation it may be necessary to install additional measures such as graphic, aural and tactile way-finding and exit sign systems. These measures may also be included as part of the access requirements for DDA.

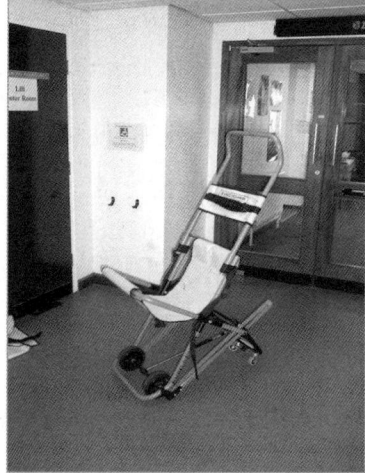

PERSONAL EMERGENCY EVACUATION PLAN (PEEP)

Figure 4-21: 'Evac+chair'. *Source: FST.*

Where an individual works in a building it is important that their needs are properly identified and adequate written arrangements developed for their safe evacuation. These arrangements may include the creation of a 'Personal Emergency Evacuation Plan' that identify the needs of a specific individual and details of other people who would assist them in an evacuation.

Generic emergency evacuation plan (GEEP)

In a building where it is probable that people will be present as visitors who will need assistance with evacuation a similar generic system should be devised and implemented.

4.3 - Fire detection and fire alarms

Common fire detection and alarm systems

TYPES OF AUTOMATIC FIRE DETECTION

The majority of serious fires occur at night when people are not present to deal with them. In the day time many large fires start in parts of buildings, such as store rooms, which are infrequently visited. In these situations it is dangerous to rely solely on people to detect fire because they may either not be there at the crucial moment or may well react incorrectly. The most effective precaution against delay in a fire being discovered and the alarm being raised is to install an automatic detection system.

The purpose of an automatic detection system is to ensure that in the event of a fire occupants are warned, so that they can be evacuated at an early stage and to ensure that the Fire and Rescue Service arrives at the premises before the fire has got out of control.

There are opportunities to detect fires at the four stages of a fire:

1. Invisible products of combustion, including carbon monoxide detectors.
2. Visible smoke.
3. Flame.
4. Heat.

The methods of detection tend to reflect these opportunities. Automatic detection systems should be installed by specialist firms and normally conform to British Standard 5839.

Fire detection equipment

The following list shows the types of fire detection equipment which have been used and which may be used to detect fire. Fire detection is normally carried out by one of five ways: -

- **Spot detectors -** static detector covering certain size floor area.
- **Line detector -** linear heat detector cable that can be laid around an area to give protection.
- **Beam detector -** beam of light (normally Infra-red) covering a large floor area.
- **Sampling detectors -** range of pipe work connecting different areas back to a detector head. The air from each area can be sampled in turn.
- **Scanning detectors -** moving detector which sweeps a large area.

Smoke detection

There are different types of smoke detectors, the main ones being optical/multi-sensor, ionisation and beam detectors.

Optical smoke detectors operate on the principle of infra red light refracting off smoke particles entering the detection chamber. This makes this type of detector more sensitive to smouldering fires such as modern fabrics or furnishings.

Optical detectors are more prone to false alarms caused by steam, found in some textile processes, or dusty environments such as those arising from building works.

Beam detectors comprise a transmitter and receiver. The transmitter emits an infra red beam from the transmitter (TX) to the receiver (RX), and the beam detects obscuration by smoke.

Ionisation detectors operate on the principle of charred smoke particles passing between two electrodes causing a small current flow. This makes this type of detector more suitable for fast flaming fires such as those involving paper or wood. Ionisation detectors are more prone to false alarms from burning odours, i.e., those smelt outside a kitchen when food is overcooked and burnt.

Heat detection

There are two main methods of operation for heat detectors - rate of rise and fixed temperature - and the majority of modern heat detectors will in fact respond to either.

- *Rate of rise heat detectors* will respond to a rapid increase in temperature. Rate of rise detectors are most suitable for areas where a smoke detector is undesirable, for example, a staff room where smoking is allowed.
- *Fixed temperature heat detectors* have a sensing element fixed at a particular temperature. When this is reached, the detector operates. Fixed temperature heat detectors are ideal for kitchens or boiler rooms where a rate of rise heat detector would be unsuitable because heat is part of the process.

Photo thermal detectors analyse both change in temperature as well as density of smoke or smoke-like phenomena. This can considerably reduce the potential for false alarms.

CATEGORIES OF FIRE ALARM AND DETECTION SYSTEMS AND THEIR OBJECTIVES (BS 5839, PART 1)

Property risk/protection

In this situation the objective is to summon the Fire and Rescue Service in the early stages of a fire.

- *Type P1:* Property protection, automatic detection installed throughout the protected building.
- *Type P2:* Property protection, automatic detection in designated areas.

Life risk/protection

In this situation the objective is to protect people from loss of life or injury.

- *Type M:* Manual system (call points).
- *Type L5:* Life safety generally when specific fire engineering solutions or where P1 insurance is required.
- *Type L4:* Life safety system, same as a manual system plus smoke detection on escape route.
- *Type L3:* Life safety system, same as a manual system plus smoke detection on escape route and heat or smoke detection in adjacent rooms.
- *Type L2:* Life safety system, same as L3 but detection in fire hazard/risk of ignition i.e. kitchens, sleeping areas and other specified areas.
- *Type L1:* Life safety system, similar to P1 but the audibility is more critical.

FIRE ALARM ZONING

Fire alarm zones are essentially a convenient way of dividing up a building to assist in quickly locating the position of a fire.

The zone boundaries are not physical features of the building, although it is normal to make the zone boundary coincide with walls, floors and specifically fire compartments.

When splitting the areas covered by the fire alarm system into Zones, certain basic rules are applied:

- The floor area of a single zone should not exceed 2,000m^2.
- Two faults should not remove protection from an area greater than 10,000m^2 (for addressable systems). An addressable system is one that gives unique identification to the actuating device detector/call point which has been activated.
- If the total floor area of the building is 300m^2 or less then it may be regarded as a single zone.
- If the total floor area exceeds 300m^2 then all zones should be restricted to a single floor level.
- As an exception to the above, stairwells, lift shafts or other vertical shafts (non stop risers) within a single fire compartment should be considered as one or more separate zones.
- The maximum distance travelled within a zone to locate the fire should not exceed 60 metres.

ALARM SIGNALLING

Warning devices

The object of the fire alarm system is to warn occupants of a fire situation. The normal method of achieving this is by the use of audible warning devices such as bells or sirens. Any warning device should be audible over background noise levels in all areas. They should also be discernible from other alarms or sounds so as not to cause confusion or complacency. In noisy areas or if audibility is a concern due to hearing impairments or other factors, then visual devices may be necessary.

If visual warnings are used, their location and type of operation is critical and must ensure that they will be seen even if workers are engrossed in a specific task. Devices which vibrate to bring the alarm to peoples' attention may also be useful for providing warning to a hearing impaired person. Voice sounders can also be used to provide an initial audible alarm, followed by a pre-recorded or voice message.

- One sounder should be located near the control panel or entrance on a separate circuit. Addressable systems should be wired from the control to a sounder protected by a short circuit isolator.
- All the sounders should sound similar to avoid confusion.
- A minimum of 65 dB is required in general areas or 5 dB above any background noise which persists for more than 30 seconds.
- Where high noise levels exist visual indication such as strobes may be required.
- Where sleeping people are to be woken then 75 dB is required at the bedhead.
- A loss of 30 dB per door should be allowed for. To ensure 75 dB at the bedhead a sounder per bedroom is recommended.
- For areas where there are people with impaired hearing the approval of devices for people with impaired hearing would be the subject of consultation with the users.

Single stage alarm

The alarm sounds throughout the whole of the building and calls for total evacuation.

Two-stage alarm

In certain large/high rise buildings, it may be better to evacuate first the areas of high risk, usually those closest to the fire or immediately above it. In this case, an evacuation signal is given in the restricted area, together with an alert signal in other areas. If this type of system is required, early consultation with the Fire Authority is essential.

USE OF ALARM RECEIVING CENTRES

An alarm receiving centre is a permanently manned centre which is usually provided by a commercial organisation. The operators of the centre, upon receipt of a fire signal, notify the fire service. As future policies are introduced by the fire and rescue services to help manage false alarms, companies with fire alarm systems which are remotely monitored by alarm receiving centres will need to be registered, and as such the company will need to comply with standards and demonstrate good fire management.

MANUAL AND AUTOMATIC SYSTEMS

The simplest type of fire alarm systems is a manually operated system, for example a hand bell or whistle. Their limitation is the size of building in which it is possible to hear it and the need to be located conveniently. Some are portable and could, therefore, be prone to loss or theft.

Stand alone call points which operate a local alarm are now available. They offer an improvement on the traditional manual alarm as the individual does not have to remain in the area once the alarm has been operated. As with manual alarms audibility of any one alarm must be sufficient to give warning throughout the building or workplace.

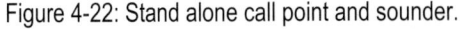

Figure 4-22: Stand alone call point and sounder.   *Source: FST.*   Figure 4-23: Fire alarm call point.   *Source: ACT.*

Call points

The fire alarm call point is the standard method of operation for a fire alarm system. However it should be ensured that workers know how to operate the system to avoid error in use or delay because of concern about how to use it. " Will I cut my finger?" and "Where's the break glass hammer?" are typical questions that workers will ask. Call points should be sited so that they are visible and, if necessary, signage may be required to ensure that they can be located. Call points are normally positioned 1.4m above floor level, on escape routes. The maximum distance a person has to travel to reach a call point should not normally exceed 30m (direct distance) or 45m (actual distance).

Selection of fire detection and fire alarm systems

LIFE RISK/PROTECTION

If it is decided that a fire alarm system is required in a premises for life risk/protection reasons, then one of the Type L systems listed above should be installed. The outcome of the fire risk assessment should decide the standard and type of alarm that is required. The traditional types that are often applied are the Type L1 and L3 standard.

PROPERTY RISK/PROTECTION

If an alarm is only being installed to provide protection for a property risk, then a Type P system may be sufficient. If the risk was specific to one or two areas, then a Type P2 system may be sufficient. If installing this type of system care would need to be taken to ensure that the property risk does not also present a life safety risk.

BEHAVIOURAL ISSUES

It is a well recorded fact that individual people frequently do not respond as they should to a traditional fire alarm system evacuation signal. They often fail to evacuate immediately in an orderly manner. As part of the risk assessment the expected behaviour of people on hearing an alarm should be modelled and the results built into the design of the fire evacuation system. In anticipation of behavioural issues it may be necessary to appoint Fire Marshals who would identify the effects of these behavioural issues at the time of alarm and encourage smooth and timely evacuation. Practising evacuation routines by holding a periodic fire drill can assist in re-enforcing the correct behaviour.

It should be borne in mind that if false alarms prevail they will tend to have a negative effect on behaviour and will re-inforce a tendency not to respond to alarms. It is therefore important to minimise false alarms.

SOCIAL BEHAVIOUR

Linked to the individual behavioural issues is the effect of social behaviour or peer pressure at the time the alarm is sounded. Again the extent to which this is a factor needs to be determined and accounted for. In some premises an immediate total evacuation may not be desirable, for example, in night clubs, shops, theatres, cinemas. A controlled evacuation by the staff may be preferred in order to prevent distress and panic to the occupants. In this situation the alarm is communicated initially only to staff, sometimes in coded announcements or by radio communication methods. When the alarm is communicated to staff they move the occupants towards the exits in a progressive, controlled manner. Such a system should only be used where there are sufficient members of staff and they have been fully trained in their actions in the event of fire and evacuation methods. The premises would still need to have available a 'standard alarm system' for use in situations where staff levels are low or the controlled evacuation is no longer sustainable.

MINIMISING FALSE ALARMS

False alarms have the potential to cause substantial disruption to the smooth running of a business and in addition place a tremendous burden on the fire and rescue service resources. Regular false alarms can cause building users to disregard alarm signals leading to incorrect actions in the event of a real fire situation. False alarms can broadly be divided into four categories:

- Unwanted alarms.
- Equipment false alarms.
- Malicious false alarms.
- False alarms with good intent.

Unwanted alarms are those that are caused by a combination of factors. Environmental conditions, including fire-like phenomena such as steam, aerosol spray or dust, or inappropriate action by people in the building, such as smoking, may trigger smoke detectors. The following is designed to assist with selection of equipment to avoid common potential unwanted alarm conditions: BS 5839, Part 1 Fire Detection and alarm systems for buildings, code of practice for the design, installation and servicing (BS 5839) gives comprehensive guidance on the subject and should be consulted for in depth guidance.

Room	Equipment
Kitchens	Smoke detectors should never be used
Areas close to kitchens	Avoid rate of rise heat detectors Avoid smoke detectors if possible Do not install ionisation smoke detectors Install photo thermal detector
Rooms in which toasters are used	Avoid smoke detectors if possible Do not install ionisation smoke detectors Install photo thermal detector

Rooms in which people smoke	Avoid smoke detectors if possible
	Do not install optical smoke detectors
	Install photo thermal detector
Bathrooms, shower rooms and areas	Avoid smoke detectors if possible where steam occurs
	Do not install optical smoke detectors
	Install photo thermal detector
Areas with high dust concentrations	Avoid smoke detectors if possible
	Do not install optical smoke detectors
	Install photo thermal detector
Areas where the sensing element is situated	Do not install ionisation smoke detectors subject to high air velocity
Areas with exhaust fumes	Avoid smoke detectors if possible
	Do not install ionisation smoke detectors
	Do not install beam detectors
	Install photo thermal detector
Areas close to openable windows	Avoid smoke detectors if possible
	Do not install ionisation smoke detectors

Figure 4-24: Minimising fire alarms.   *Source: BS 5839.*

When installing fire detection into buildings the equipment should be designed and selected to provide the fastest detection response time and the minimum number of false alarms. The number of false alarms which are created by fire detection systems and attended by the British Fire and Rescue Service has increased dramatically over recent times. The fire and rescue service attended approximately 450,000 false alarms in 2004 which equates to 50% of all calls they receive. Of these calls 64% were false alarms due to equipment design or malfunction.

The costs to the government of this are considered to be unsustainable and as a result a new policy on Fire and Rescue Service responses has been introduced. All remotely manned fire alarm systems (RMFAS) will need to be registered with the fire and rescue service. A calculation will be made on the number of false alarms a given alarm system will be allowed to generate, based on the number of automatic detectors on the system. Depending upon the result the system and therefore the premises it protects will be placed in one of three levels of response.

- **Attendance level one -** An immediate emergency response with an attendance based on a risk assessment of operational needs, not less than one fire appliance.
- **Attendance level two -** If no confirmation call is received attendance is as above, but attendance is made under non-emergency conditions.
- **Attendance level three -** No emergency response is made until a confirmation of fire is received via 999 or 112 systems. Confirmation of fire will then result in an Attendance Level One response as above.

If an organisation's fire alarm system gets taken off cover it will be necessary to:

- Revise the risk assessment and fire safety management arrangements due to changes in response to building.
- Inform your insurance company of changes.

Furthermore, in order to become reinstated to level one cover it will be necessary to demonstrate the following:

- Demonstrate measures are in place to reduce false activations.
- Demonstrate competence as a responsible person as described in BS 5839.

The responsible person may be asked by the fire and rescue service to provide a written improvement plan.

REQUIREMENTS OF DISABLED PERSONS

Consideration to people with disabilities needs to be made. The standard mounting height for alarm call points is 1.4m from floor level. This may need to be lowered to accommodate wheelchair users. In addition, the usual maximum distances between call points may need to be reduced if it can reasonably be assumed that the first person to raise the alarm may have a disability that that would prevent them getting to call points easily.

See also - requirements for means of escape for disabled people - earlier in this element.

Requirements for maintenance and testing of fire alarm systems

Article 17 of the RRFSO sets out a requirement that "equipment and devices are subject to a suitable system of maintenance and are maintained in an efficient state, in efficient working order and in good repair". In order to meet this requirement it is essential that a planned inspection and test regime be implemented. One such regime would include the following.

Daily - Check to see if the system is indicating a fault and that any corrective actions have taken place.

Weekly - Test the system by operating a manual call point (different one each week).

Periodic Inspection - Subject to risk assessment, but should not exceed 6 months between inspections. Check the system log and ensure that corrective actions have taken place. Visually inspect all items of equipment, to ensure that the system is not obstructed or rendered inappropriate by change of use. Check for any false alarms, compare to nationally accepted levels and take appropriate action if unacceptable.

Test the system on standby power to ensure that the battery is functioning correctly. Check all outputs for correct operation. Check all controls and indicators. Check remote signalling equipment.

Additionally any other special checks should be made - for example beam detectors for correct alignment.

Over 12 month period - Carried out at least 2 inspections.

In addition to the periodic inspection: Over and above the standard periodic inspections the following must be completed during a 12 month period: Test all manual call points and fire detectors for their correct operation. Inspect the analogue detector levels to ensure that they are within correct levels. Check all alarm devices for correct operation. Visually inspect all accessible cable fixings. Confirm the cause and effect programming is correct and up to date.

Figure 4-25: Good management of testing.   *Source: FST.*

4.4 - Means of fighting fire

Portable fire-fighting systems and equipment

CLASSIFICATION OF FIRES

A basic understanding of the classes of fire needs to be understood in order to provide and use the correct fire fighting system and equipment for the likely fires that will occur in premises. In addition many portable fire-fighting equipment (fire extinguishers) state the classes of fire for which they are suitable. There are 5 classes into which fires can fall:

CLASS A		Fire involving solids (wood, paper, plastics, etc., usually of an organic nature).
CLASS B		Fires involving liquids or liquefiable solids (petrol, oil, paint, fat, wax, etc.).
CLASS C		Fires involving gases (liquefied petroleum gas, natural gas, acetylene, etc.).
CLASS D		Fires involving metals (sodium, magnesium and many metal powders, etc.).
ELECTRICAL HAZARDS		Although not a true class of fire, we should also consider fires in electrical equipment.
CLASS F		Fire involving cooking fats/oils.

Figure 4-26: Classification of fires.   *Source: Rivington.*

EXTINGUISHING MEDIA AND MODE OF ACTION

Portable fire fighting equipment (fire extinguishers) - introduction

The familiar coding of the whole body of an extinguisher in a single discernible colour has now disappeared under the new British/European standard BS EN 3. From 1st January 1997 all types of new certified fire extinguishers should have a red body like the existing water-type extinguisher. In addition, BS EN 3 allows manufacturers to use up to 5% of the extinguisher casing in another colour in order to differentiate between extinguishers that use a different extinguishing medium. A new British Standard (BS 7863) recommends that manufacturers affix different colour coded panels (for example, labels or bands) using the existing colour code scheme noted below when describing the different extinguishers. The changes introduced by BS EN 3 are not legal requirements and existing extinguishers do not have to be replaced, but may be replaced as they become unserviceable.

Portable fire fighting equipment (fire extinguishers) - use

Water

Water extinguishers should only be used on Class A fires - those involving solids like paper and wood. Water works by cooling the burning material to below its ignition temperature, therefore removing the heat part of the fire triangle causing the fire to go out. In addition to the cooling effect, steam is produced in the fire area due to the effects of heat on the water; this aids in the extinguishing process by tending to smother the fire. Water is the most common form of extinguishing media and can be used on the majority of fires involving solid materials. However it *must not be used on liquid fires or in the vicinity of live electrical equipment* as it can cause liquid fires to spread and there is a risk of electrocution with live electrical equipment.

Foam

Foam is especially useful for extinguishing Class B fires - those involving burning liquids and solids which melt and turn to liquids as they burn. Foam works in several ways to extinguish the fire, the main way being to smother the burning liquid, i.e. to stop the oxygen reaching the fire zone.

Foam can also be used to prevent flammable vapours escaping from spilled volatile liquids.

Modern spray foam extinguishers are ideally suited to Class A fires, and may often be used in preference to water type extinguishers.

A new style of 'wet chemical' extinguishers has been designed to specifically deal with the type of risk presented by Class F fires - cooking fats and oils. The foam congeals and cuts off the oxygen supply to the fire.

However, as water is one of the constituent parts of the foam it *must not be used in the vicinity of live electrical equipment, unless it has passed an electrical conductivity test and then only with great care from a distance of more than one metre.*

Powder (dry powder)

One of the ways in which powder works to extinguish a fire is the smothering effect, whereby it forms a thin film of powder on the burning liquid thus excluding air. It also chemically interferes with the flame propagation process, which makes powder excellent for the rapid knock down (flame suppression) of fires involving large flammable liquid spills.

Powders generally provide extinction faster than foam, but there is a greater risk of re-ignition and this should always be borne in mind. The high performance powders now on the market can be used on Class A fires, but the normal powders will only subdue this type of fire for a short while.

If used indoors, a powder extinguisher can cause problems for the operator due to the inhalation of the powder and obscuration of vision. It is therefore imperative that the exit route is clear and available to those using the extinguisher before it is operated.

Powder extinguishing media may be used on live electrical equipment safely, but may cause damage to it.

Special powders have been developed to deal with Class D - metal fires, e.g. aluminium swarf. The powder inside the extinguisher may vary depending upon the metal risks involved, but may include pyromet, graphite and talc or salt. It may also be possible to deal with metal fires by using a supply of dry sand to smother the fire.

Carbon dioxide (CO_2)

Carbon dioxide replaces the oxygen in the atmosphere surrounding the fuel and the fire is extinguished.

As most carbon dioxide extinguishers last only a few seconds, only small fires should be tackled with this type of extinguisher. Because it replaces the oxygen, CO_2 is an asphyxiant and should not be used in very small, confined space unless the operator can withdraw quickly. Unless the fire is completely extinct, it will take hold again as soon as the CO_2 disperses, dispersal usually occurs within a few seconds.

CO_2 is safe and excellent for use on live electrical equipment. It may be used on small Class B fires in their early stages, indoors or outdoors, provided there is little air movement. Carbon dioxide is not regarded generally as the best choice for Class B fires, cooling is limited and such a fire may reignite.

When working correctly a CO_2 extinguisher is very noisy due to the rapid expansion of gas. This expansion causes severe cooling around the discharge horn and can freeze skin which is in contact with uninsulated parts of the horn.

Isolation of fuel

Except in very small occurrences, a Class C fire involving flammable gases should not normally be extinguished, and should only be extinguished if the gas supply can be isolated.

If a leak from a gas cylinder or pipework is burning, the danger area can be identified, i.e. the flames can be seen, and anything which is being affected by the flame can be seen. Where possible the area around the fire should be protected until the leak can be stopped at source, by closing the valve, etc.

This removes the third side of the fire triangle, the fuel. Any remaining small Class A or B fires can be dealt with using the correct extinguisher.

If, however, the fire is extinguished before shutting off the supply, the area of danger cannot then be seen. The possibility of an explosion due to a highly flammable gas cloud spreading throughout the area then becomes a significant risk.

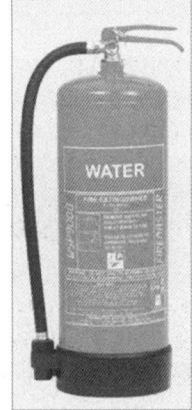
Water - red and white

Foam - red and cream

Powder - red and blue

Figure 4-27: Various fire extinguishing media.

Source: FST.

Carbon dioxide - red and black

Wet chemical - red and yellow

Fire blanket - red and white

Figure 4-28: Various fire extinguishing media.

Source: FST.

Summary matrix - fire extinguishers

MEDIA	COLOUR CODE	METHOD	CLASS 'A'	CLASS 'B'	CLASS 'C'	CLASS 'D'	ELECTRIC	CLASS 'F'
WATER	Red	Cools	Yes	No	No	No	No	No
SPRAY FOAM	Cream	Smothers	Yes	Yes	No	No	No	Special Forms
DRY POWDER	Blue	Smothers & Chemical	Limited	Yes	Yes & Isolate	Special Powders	Yes - Low Voltage	No
HALON	Green	Chemical & Smothers	ILLEGAL					
CARBON DIOXIDE	Black	Smothers	No	Yes - Small Fires	No	No	Yes	No

Figure 4-29: Summary matrix - fire extinguishers

Source: ACT.

Portable fire fighting equipment (fire extinguishers) - limitations

Fire classification

Fire extinguishers are limited firstly to the class of fire on which they can be successfully used. It would not be good practice and may indeed be dangerous to use extinguishers for types of fires for which they are not approved.

Duration of discharge

The effectiveness of a fire extinguisher is then limited by the duration of discharge. Some extinguishers only last for very small periods of time and for this reason can only be used on small fires.

The minimum duration of discharge for extinguishers is as shown in the table below:

Nominal charge of extinguisher Kg or litres	Minimum duration of discharge of extinguishers (in seconds).
Up to and including 3	6
More than 3 but less than or equal to 6	9
More than 6 but less than or equal to 10	12
More than 10	15

Figure 4-30: Minimum duration of discharge. *Source: BS 5306.*

Range of discharge

The extinguisher is also limited by the range (or throw) of the discharge. This would affect the maximum distance that a person could be from a fire and still be effective in extinguishing the fire. Since it would not be safe for the operator of an extinguisher to get too close to fires, if the range of discharge is too short, then it will not be possible to extinguish the fire using this type of equipment.

SITING, INSTALLATION, MAINTENANCE AND TRAINING REQUIREMENTS

Portable fire fighting equipment (fire extinguishers) - siting and installation

Portable fire extinguishers should always be sited:

- On the line of escape routes.
- Near, but not too near, to danger points.
- Near to room exits inside or outside according to occupancy and/or risk.
- In multi-storey buildings, at same position on each floor, i.e. top of stair flights or at corners in corridors.
- Where possible in groups forming fire points.
- Where possible in shallow recesses, if sited on a wall.
- So that no person need travel more than 30 metres to reach an extinguisher.
- With the carrying handle about one metre from the floor to facilitate ease of handling, and removal from wall bracket, or on purpose designed floor stands.
- Away from excesses of heat of cold.

Figure 4-31: Extinguisher well sited on escape route. *Source FST.*   Figure 4-32: Incorrectly sited plant. *Source: FST.*

Fire Ratings of extinguishers

Some attempt is made at showing the limits of a fire extinguisher by stating its fire rating on it. However, this is only done for Class A and Class B fires.

Class A fire ratings are achieved by an extinguisher successfully putting out a designated size of test fire. The test fire is made from a particular type of wood Pinus Silvestris and the rating achieved relates directly to the length and number of pieces of timber used to construct the test fire, e.g. a 13A rated extinguisher, extinguished a fire constructed of 14 layers of sticks, each transverse layer consisting of 13 sticks (500 mm each), the test fire being 1.3 metres long.

Class B fire ratings are similarly achieved but the fires consist of specified containers of flammable liquid (aliphatic hydrocarbon). The test fire rating achieved relates directly to the volume in litres of fuel, e.g. a 55B rated extinguisher extinguished a fire consisting of 55 litres of fuel with a surface area of 1.73 square metres.

As can be seen from the illustrations given above, the procedure for test fire ratings is quite involved and, if required, further guidance on this matter should be sought from BS EN 3-7:2004 Portable Fire Extinguishers: characteristics, performance requirements and testing.

The number of fire extinguishers required per floor/area can be worked out by reference to BS 5306: pt 8 - Fire extinguishing installations and equipment on premises.

Class A

You must provide a minimum number of fire fighting equipment for a class A fire to comply with BS5306: pt 8 - Fire extinguishing installations and equipment on premises.

- The basic provision for extinguishers is a minimum of 2 extinguishers per floor.
- The total Class A ratings worth of extinguishers should be no less than 0.065 x floor area of storey (m^2), but with a minimum rating of 26A.
- However in small single occupancy buildings with an upper floor area not exceeding 100m^2 a minimum rating of 13A may be sufficient.

Class A Fire Rating required = 0.065 x floor area (m^2).

Class B

The provision of extinguishers for a Class B risk is unfortunately more complicated. The workplace needs to be assessed as follows:

- Each room or enclosure should be considered separately.
- Fire risks more than 20 metres apart should be considered separately.
- Fire risks sited within 20 metres of another fire risk should be assessed as either an undivided or a divided group.

Contained fires

Single open top containers

The minimum requirement for Class B rating can be obtained from the table below. The surface area of the container is used to determine the rating required.

Undivided group of containers

This applies to containers less than 2m apart. The total surface area of all containers in the group is used to determine the recommended rating.

Divided group of containers

Containers more than 2m but less than 20m apart fall into this group. The surface area of the largest container (or aggregate surface area of the largest individual group) or one-third of the aggregate surface area of all the containers in the group, whichever is the greater, is used to calculate the rating required.

Maximum area of Class B fire (deep liquid) for which extinguishers are suitable			
Extinguisher Rating	*Maximum area for three extinguishers (foam extinguishers only)*	*Maximum area for two extinguishers*	*Maximum area for one extinguisher*
	m^2	m^2	m^2
21B	0.42	0.26	0.14
34B	0.68	0.42	0.23
55B	1.10	0.69	0.37
70B	1.40	0.88	0.47
89B	1,78	1.11	0.59
113B	2.26	1.41	0.75
144B	2.88	1.8	0.96
183B	3.66	2.29	1.22
233B	4.66	2.91	1.55

Figure 4-33: Maximum area of Class B fire.   *Source: BS 5306, part 8.*

Spillage fires

A Class B fire rating equal to 10 x volume (litres) of possible spillage is required.

Portable fire fighting equipment (fire extinguishers) - maintenance

British Standard Code of Practice BS 5306: pt 3 - Fire extinguishing installations and equipment on premises details the inspection, maintenance and testing of portable fire extinguishers as follows:

(a)   Monthly inspection

A monthly check should be carried out to ensure that extinguishers are in their proper place and have not been discharged, lost pressure or suffered obvious damage. This check is recommended as weekly in BS5306; pt 3 - Fire extinguishing installations and equipment on premises.

(b)   Annual inspection and maintenance

A more thorough inspection of extinguishers, spare gas cartridges and replacement charges should be carried out by a competent person on an annual basis. This may include internal and external inspection dependent upon the type of extinguisher.

(c)   Test by discharge

Extinguishers should be tested by discharge at intervals as detailed below. The time interval should be taken from the date of manufacture or the last actual discharge.

Extinguisher type	Interval of discharge
Water (stored pressure)	Every 5 years
Foam (all types)	Every 5 years
Water (gas cartridge)	Every 5 years
Powder (gas cartridge)	Every 5 years
Powder (stored pressure valve operated)	Every 5 years
Carbon Dioxide (all types)	Every 10 years

Figure 4-34: Extinguisher intervals of discharge.   *Source: BS 5306, part 3.*

The above information is only a portion of the information on maintenance, inspections and testing. For further details the British Standard should be consulted.

Portable fire-fighting equipment (fire extinguishers) - training

Though the RRFSO does not specify specific training for fire extinguishers it does require "suitable and sufficient instruction and training on the appropriate precautions and actions to be taken by employee". It further requires that training be carried out periodically where appropriate. If an employee is expected to make use of extinguishers this Article confirms that initial and periodic training must take place. Any person, who may be called upon to use a fire extinguisher, should be trained in the selection and practical use of the equipment.

Fixed fire-fighting systems and equipment

HOSEREELS

Hosereels - use

Use on fire and effect

Hosereels are designed for use on Class A - Carbonaceous fires. The hosereel acts as a replacement for water fire extinguishers and it is said that one hosereel equates to 4 x 9 litre water extinguishers.

Modern hosereels have an adjustable nozzle and can be adjusted to give a jet of water, water spray or a combination of both. The water jet is normally used for its "striking power" in attacking the seat of a fire. The jet of water should be "played" across the fire surface and into the heart of the fire to extinguish embers, etc. The spray pattern setting has less pressure at point of contact with the burning material and as such is less prone to spreading burning material than a single water jet setting. The water spray can therefore be used if the burning material is easily disturbed with the possibility of spreading the fire. The spray pattern produced allows larger areas to be covered in one go rather than if the water jet is used.

The water spray can also be used for protection purposes by placing a curtain of water droplets between the fire and the person operating the hosereel and therefore providing a barrier which reduces the radiated heat that is absorbed by the operator.

Hosereels - advantages/disadvantages

Advantages

- Continuous supply of water - no time constraint.
- Greater quantity of water is delivered than from an extinguisher which has a better effect at extinguishing the fire.
- The person who is attempting to extinguish the fire does not need to get as close as when using extinguishers.
- A spray pattern can be produced to protect the user from radiated heat.

Disadvantages

- Considerable physical effort may be required to pull the hosereel to the fire, especially if the route is twisting and has lots of obstructions.
- Any doors through which the hosereel is pulled will become wedged open by the reel. This will present the possible problem of smoke travel within the building.
- Can only be used on Class 'A' Carbonaceous material fires.
- May create a trip hazard on the escape route.
- People may stay in the fire area for too long, and put themselves at excessive risk.

Hosereels - limitations on use

Hosereels should be connected to a permanent water supply so they are not limited by discharge time factors as are extinguishers.

New hosereel, when manufactured, should comply with EN 671-3:2000: Fixed fire fighting systems, which permits up to a maximum of 30 metres of hosereel in one piece. However, equipment purchased and installed prior to EN 671 being applicable would have been manufactured to BS 5306 Part 1 Fire extinguishing installations and equipment on premises. This allowed up to 45m on one drum of hosereel and equipment manufactured to this standard may remain in use till it is replaced. Each of these standards limits the range for use of any one hosereel by virtue of the distance that it is located away from a fire and the length of the hose.

One additional limiting factor to the use of hosereels is the friction drag created by pulling the hosereel along the floor. Considerable physical strength is required to pull 30 metres of hose, and for this reason consideration should be given to shorter lengths of hosereel being located at more frequent intervals.

Another limiting factor is the likely route through which the hosereel will need to pass. If this route includes a lot of corners, turns or doors, then hosereels may not be the most suitable fire-fighting equipment to be provided. A point to remember is that every door between the 'fire room' and the main escape route will be wedged open by the hosereel and may allow smoke and hot gases to make their way to other parts of the building

Hosereels - siting

- One hosereel should be provided for every 800m^2 or part thereof.
- Hosereels should be sited in prominent and accessible positions at each floor level adjacent to exits in corridors on exit routes in such a way that the nozzle can be taken into every room and every part of the room.
- Preferably hosereels should be installed into recesses so as not to obstruct the means of escape.
- Any doors fitted to hosereel cupboard doors should open through 180 degrees.

Hosereels - maintenance

(a) Monthly inspection

A check should be made to ensure that the hosereel system is intact and has no obvious signs of damage. The operating valves should be checked to ensure that they are in the correct position and that they are free moving.

(b) Annual test

- The hosereel should be run out to its full length and checked for damage or defects.
- The hosereel valve and the nozzle should be checked to ensure free movement throughout its operating range.
- The water supply should be switched on and the nozzle operated to check that a suitable water jet is obtained.
- It would be good practice to check that 24 litres per minute of water is supplied from the topmost reel - with the 2 topmost reels operating at the same time.
- Whilst the hosereel is pressurised, a check should be made for any leaks from the system.

Hosereels - training

Though the Regulatory Reform (Fire Safety) Order (RRFSO) 2005 does not specify specific training for fire extinguishers it does require "suitable and sufficient instruction and training on the appropriate precautions and actions to be taken by employee". It further requires that training be carried out periodically where appropriate. If an employee is expected to make use of hosereels this Article confirms that initial and periodic training must take place. The need for training of employees will increase if hosereels are installed in buildings as people may be tempted to stay in buildings for too long attempting to fight the fire. Consideration should be given to using teams of two or more to operate hosereels to help assure an individual's safety.

AUTOMATIC SPRINKLER SYSTEMS

Automatic sprinkler systems are used more than any other fixed fire protection system and over 40 million sprinklers are fitted worldwide each year.

Sprinkler systems have been proven in use for well over 100 years. Possibly the oldest in Britain was fitted in 1812 at the Theatre Royal Drury Lane and in updated form is still in use today.

All areas of the building to be protected are covered by a grid of pipes with sprinkler heads fitted into them at regular intervals. Water from a tank via pumps or from the town main (if it can give enough flow) fills the pipes.

Each sprinkler head will open when it reaches a specific temperature and spray water on to a fire. The hot gases from a fire are usually enough to make it operate. Only the sprinklers over the fire open. The others remain closed. This limits any damage to areas where there is no fire and reduces the amount of water needed. The sprinkler heads are spaced, generally on the ceiling, so that if one or more operate there is always sufficient flow of water. The flow is calculated so that there is always enough to control a fire taking into account the size and construction of the building and the goods stored in it or its use.

Sprinkler heads can be placed in enclosed roof spaces and into floor ducts to protect areas where a fire can start without being noticed. In a large warehouse sprinklers may be placed in the storage racks as well as at the roof.

At the point where the water enters the sprinkler system there is a valve. This can be used to shut off the system for maintenance. For safety reasons it is kept locked open and only authorised persons should be able to close it. If a sprinkler opens and water flows through the valve it lets water into another pipe that causes a bell to ring. In this way the sprinkler system both controls the fire and gives an alarm using water, not electricity.

Sprinkler types

Wet pipe

These are the most common systems and are used in buildings where there is no risk of freezing. They are fast to react because water is always in the pipes above the sprinkler heads.

Wet systems are required for multi-storey or high rise buildings and for life safety.

Alternate

As the name suggests alternate systems can have the pipes full of water for the summer and be drained down and filled with air (under pressure) for the winter. This is important for buildings that are not heated.

Dry pipe

The pipes are filled with air under pressure at all times and the water is held back by the control valve. When a sprinkler head opens the drop in air pressure opens the valve and water flows into the pipe work and on to the fire. Dry pipe systems are used where wet or alternate systems cannot be used.

Pre-action

Like dry pipe systems the pipes are filled with air but water is only let into pipes when the detector, for example, smoke detectors, operates. Pre-action systems are used where it is not acceptable to have the pipes full of water unless there is a fire.

Deluge and re-cycling installations

These are not strictly sprinkler systems and are only used in special cases for industrial risks.

Features of a sprinkler system

Controlling valves (wet system) - sprinkler stop valve

This valve is of the usual wedge type and is provided on the supply side as a means of cutting off the water supply to the installation following the extinguishment of the fire and is normally kept strapped and padlocked in the open position, a strap being used so that in the event of the key not being available to open the lock, the strap can be cut thus enabling the stop valve to be closed. Upon arrival at a fire or where a sprinkler system has actuated, the officer in charge of the first attendance will station a firefighter at the stop valve to ensure that it is **not** closed down until he/she gives orders to that effect.

Pressure gauges

Two pressure gauges are fitted to each set of valves; one is connected to the supply side of the stop valve and the other to the delivery or installation side of the valve. The former indicates the pressure in the mains or other sources of supply, the latter to record the pressure in the installation. Also in the case of the latter, to record the air pressure on the installation when the system is on air, in the case of the alternative system.

Sprinkler heads

Two main types of sprinkler heads are in use throughout the country, all conforming to a similar design and principle of operation. These are fusible soldered strut or quartzoid bulb. The main components of a sprinkler head are:

- Main body with screw thread for screwing into pipework.
- Yoke, i.e. 'V' shaped casting.
- Distributor at the base of the yoke.

- Diaphragm (a flexible metal plate with a hole in the centre secured between the yoke casting and the main body).
- Glass valve (a hemispherical glass valve which seats against the diaphragm).
- Glass valve cap (a small brass cap which butts against the underside of the glass valve and is recessed to house one end of a metal strut).

Fusible soldered strut

With a fusible soldered strut sprinkler head, the strut is composed of three pieces of metal joined together with a low melting point solder. When this solder is softened by the heat of the fire, the strut falls apart and the glass valve is thrown clear. The water escapes in a solid half inch jet, impinges upon the distributor plate and is scattered in all directions in the form of a drenching spray. In order to protect against deterioration in the operation, certain corrosive resistant coatings are applied to the sprinkler by the manufacturer. It is also recommended that petroleum jelly be applied to all heads periodically. The operating temperatures at which sprinkler heads are designed to operate are identified by various colours.

Soldered strut	Colour of yoke arms
68/74°C	Uncoloured
93/100°C	White
141°C	Blue
182°C	Yellow
227°C	Red

Figure 4-35: Fusible soldered strut - operating temperatures for sprinkler heads.

Source: FST.

Quartzoid bulb type

The main structure of the head is retained but in place of the struts a barrel shaped bulb made of Quartzoid (a transparent material) of unusual strength and toughness is used. This bulb is hermetically sealed after being filled with a highly expandable liquid (coloured). When the head is exposed to a rise in temperature pressure within the bulb rises quickly, and fracture of the bulb results. These sprinkler heads are made to operate at various temperatures. In temperate climates, the usual operating temperature for normal situations is 68°C, but for use in special conditions, such as are met in drying stoves, ovens, etc. higher operating temperatures are necessary.

The Quartzoid bulb is so strong that it can stand any hydraulic pressure which may be applied to the interior of the sprinkler head. The filling in the whole range of Quartzoid bulbs cannot freeze however severe the frost may be. These sprinklers are equally suitable for the coldest as well as the hottest climate. The bulb is completely resistant to acid or any corrosive action and in any situation will preserve its effectiveness unimpaired for an indefinitely long period. The operating temperatures at which sprinkler heads are designed to operate are identified by various colours.

Sprinkler rating	Colour of bulbs
57°C	Orange
68°C	Red
79°C	Yellow
93°C	Green
141°C	Blue
182°C	Mauve
204 to 260°C	Black

Figure 4-36: Quartzoid bulb type - operating temperatures for sprinkler heads.

Source: FST.

Water supplies

The water supplies for a sprinkler system must be reliable under all conditions and adequate for the appropriate class of risk. They are divided into grades according to the number and type of water supplies available.

Testing of sprinkler systems

Detailed below is a typical testing regime.

Daily Checks

- The alarm connection to the Fire and Rescue Service or the remote manned centre should be tested daily if it is not automatically monitored.
- Check water levels and air pressures in pressure tanks if not automatically monitored.

Weekly Checks

- Check and record:
a. All water and air pressure gauge readings on installation, trunk mains and pressure tanks, and
b. All water levels in elevated private reservoirs, rivers, canals, lakes, water storage tanks and pressure tanks.

- Test each water motor alarm for 30 seconds.
- Automatic pump starting tests:
a. Check oil, fuel levels.
b. Simulate a pressure fall to operate automatic start.
c. Check pressure at which pump cuts in - is it correct?

- Check the electrolyte level and density of any lead acid batteries.
- If heating systems are fitted to prevent freezing, they should be checked.

Quarterly checks

A complete check is done normally by an engineer, plus a check should be made to ensure that there have been no structural changes or alteration of layout of contents that would impede the effectiveness of the sprinkler system.

In addition, there are checks at 6 month, 1 year, 3 year and 15 year intervals. These are normally completed by a service engineer.

WATER SPRAYS

High velocity water sprays

Effective fire protection is afforded by projectors which project a fine spray. The main difference between a water spray projector system and a sprinkler system is that the projectors are not only located overhead but may be sited all around the area of risk. They can be installed to discharge in a horizontal, or even, in certain circumstances, an upward direction. The main application is for the protection and containment of plant using or storing insulating oil, lubricating oil, fuel oils and other flammable liquids. High velocity water sprays project water at high velocity to emulsify oils and liquids at risk. *They may be used as an extinguishing system as well as for protection from fire.*

The water sprays system is characterised by the type of water projectors selected and these characteristics vary in regard to their density of flow, discharge angle, maximum spacing between projectors and maximum range from the equipment to be protected. The type, characteristics and range of projectors differ with each manufacturer, consequently projector coverage, design and configuration of systems offered will vary. Each water spray nozzle should include a thimble strainer integrated within the nozzle to prevent blockage of the nozzle exit by any debris in solution or inadvertently picked up in the pipework.

The point of projectors must not be moved once set, otherwise the area will no longer be fully protected.

Medium velocity water sprays

This system is used for protecting such risks as bulk hydrogen, chlorine and propane stores and has been used for cooling the external surface areas of bulk gas turbine fuel storage tanks. *It is not an extinguishing system* and unlike high pressure water spray systems operates at a lower pressure at the sprays. Volatile liquid fire situations cannot be extinguished by the application of medium velocity water sprays but close control of a fire can be obtained at the same time rendering adjacent plant and structures safe by the cooling effect of the water film. This form of protection also gives protection to personnel when fighting fires on this type of risk. Without it, close approach to the seat of the fire might not be possible.

HYDRANT SYSTEM

The main hydrant systems form ring mains and are for general use in protecting buildings and other such risks. The system is generally kept primed with clean water but not under pressure, with the pumps being on manual start. The water supply can be from a river or the sea. There are external manifolds into which the Fire and Rescue Service can connect their pumps to supplement the system.

Some hydrants are coupled to dry risers which are only pressurised when water is pumped into the low level inlet.

INERTING / EXTINGUISHMENT FLOOD SYSTEMS

Fire extinguishing systems which are designed and installed in buildings to detect and extinguish fires are commonly called inerting or extinguishment flood systems.

Inerting can be defined as the displacement of the atmosphere in a confined space by a non-combustible gas (such as nitrogen) to such an extent that the resulting atmosphere does not sustain combustion.

Traditional flood systems, such as those using carbon dioxide or nitrogen, where the displacement of air within the enclosure is necessary for their successful operation may be considered to be inerting systems. For systems that use specialist chemicals such as Inergen or OxyReduct the manufacturers claim that it is safe for people to be inside the area with the system active and operated. In order for there to be sufficient air to breathe the system cannot have displaced all of the air from the enclosure and works in a different way to standard inerting chemicals. It may be more useful therefore to think of these systems as extinguishment flood systems.

Carbon dioxide flooding

Fixed carbon dioxide (CO_2) installations in this country are usually of the high pressure type in which the gas is stored in liquid form in drawn steel cylinders each containing from 22 to 35 kg (50 to 80 lb) of CO_2 at a pressure of 48 - 58.5 bar depending on the atmospheric temperature. The cylinders may be used singly or may be connected to a manifold in batteries.

Characteristics of fixed CO_2 installations

Carbon dioxide is stored in the cylinders as a liquid, and when released it travels in the same form through the pipework to the discharge nozzles. In the case of high-pressure systems, a discharge horn is fitted to the nozzle to prevent the entrainment of air with gas, minimise turbulence, and reduce the high velocity of the discharge. The release of pressure allows the liquid to turn partly to 'snow' (solid CO_2) and partly to gas; the 'snow' emerges at a temperature of -79^{0}C.

Carbon dioxide extinguishes a fire principally by reducing the oxygen concentration in the vicinity of the burning material.

Operation of CO_2 systems

A single cylinder is often manually operated by a pull handle, provided either close to the cylinder or at a point outside the space protected. The handle is connected to a cable which, when pulled, withdraws a pin holding a weight in place above the operating level on the cylinder, so that the withdrawal of the pin allows the weight to drop on the lever and so open the cylinder. A battery of cylinders can be similarly operated. Above the lever of each cylinder is a weight and all the weights are held up by a common operating shaft. When the support, which keeps the shaft in place, is released (by pulling the releasing handle) the weights fall on the respective levels and so operate the cylinders.

A fixed CO_2 installation may be operated manually, as we have described above, or automatically by means of air-expansion thermostats or, more usually, by fusible links.

The systems are classified as manual or automatic according to the method of actuation, although the automatic system will also have some means of manual control. Before work or inspections are carried out in any enclosure protected by automatic CO_2, or other chemical extinguishing equipment, the automatic control is rendered inoperative and the equipment left on hand control; a notice to this effect is attached to the equipment.

The automatic control is restored immediately after the persons engaged on the work or inspections have come out of the protected enclosure. Any precautions which are taken to render the automatic control inoperative are noted on any Permit for Work issued for work in the protected enclosure. If it is necessary to enter a space flooded with CO_2 breathing apparatus must be worn because CO_2, although not toxic, is an asphyxiant and excludes oxygen from the room.

Environmentally friendly flood systems

Due to the removal of Halon systems from buildings alternatives are now readily available. There is generally no straight replacement for Halon, however an American company has now launched Fike 25 which utilises existing pipework. All other alternative systems are a complete new refit. There is a choice of inert gases such as argon, or suppression systems such as Du Pont FE13^{TM}.

Water mist systems

Water mist systems are hailed as one of the most rapid, safe and environmentally friendly ways of suppressing fires. Water mist systems can be used for Total Flooding, In-Cabinet Protection and Local Applications, and these can be combined to suit almost any scenario imaginable. In its various modes, water mist systems are ideally suited for the protection of art galleries, computer and telecommunications equipment, generating plant, robotic machinery, cable tunnels, kitchens, and many similar applications.

Fixed foam installations

Fixed foam installations are used against flammable liquids, and generally consist of foam pourers or foam-making branches fed with a supply of foam.

Figure 4-37: Cylinders for Flood System. *Source: AFE.*

The foam is normally the mechanical type foam produced by mixing foam-making compound and water and passing it through a foam-making generator. These generators are designed to mix the correct proportions of foam compound and air with the quantity of water which is flowing, to generate foam and then deliver it through pipework to the point of discharge. The foam so produced normally has an expansion ratio of approximately 8:1. High expansion foam with an expansion ratio of 1000:1 will be discussed at the end of this sub-section.

It is less easy to arrange for fixed foam installations to operate automatically than is the case with other extinguishing agents.

Not only is it vital that the system should operate immediately a fire starts, but it is equally important that it should be shut down as soon as the supply of compound or solution is exhausted; otherwise water alone will reach the foam pourers, with potentially disastrous results.

It is because of this that completely self-contained fixed installations are generally limited to relatively small risks, such as isolated indoor transformers, etc. For large risks, like oil storage tanks, the fixed installation consists of fixed piping (arranged to suit the risk) terminating in foam pourers or fixed monitors; the piping is run back to an appropriate point and terminates in a coupling, usually protected by a glass panel marked with the works foam inlet, together with an indication of the particular risk involved.

This arrangement ensures that foam can be applied where it is required; as long as foam compound and water under pressure is available to a foam generator or foam inductor at this point, a continuous supply of foam can be channelled to the required spot.

Use of high expansion foam

It has been shown that high expansion foam is a practical method of fire protection, particularly for special risk areas which are otherwise difficult to protect effectively. High expansion foam can fill an entire building in a matter of minutes; the foam is a good heat insulator and it is effective on ordinary surface fires as well as on flammable liquids.

Dry powder installations

Dry powders, in common with vaporising liquids, offer the advantage of a quick knock-down of fire, but unlike vaporising liquids, they have negligible toxic effects. Their major disadvantage is that they require a lot of clearing up once an installation has operated. Compacting of the powder is also a problem, due to heat or vibration or moist atmospheres during storage. This could present difficulties in the maintenance of the system especially after discharge when compacting could take place in valves, etc. Recently developed powers (e.g. 'Monnex') appear, however, to be free of this problem.

A dry powder installation consists of dry powder containers linked by pipework to discharge nozzles covering the areas of risk. When a fire occurs it is necessary to pressurise the powder so that it is forced through the pipework and discharge nozzles. This is usually done with CO_2. A line detector is linked to a lever which when actuated allows the head of a CO_2 cylinder to be pierced. The carbon dioxide thus released pressurises the dry powder and forces it over the protected area. Dry powder installations can usually be operated either automatically or manually.

4.5 - Access & facilities for the fire service for fire fighting

Legal requirements

Building Regulations make requirements for access and facilities for the Fire and Rescue Service. As the regulations relate to the initial design and construction of the building only it is not mandatory under the Building Regulations to maintain these requirements during the life of the building. However Article 38 of the Regulatory Reform Fire Safety Order requires: "Where necessary in order to safeguard the safety of fire-fighters in the event of a fire, the responsible person must ensure that the premises and any facilities, equipment and devices provided in respect of the premises for the use by or protection of fire-fighters under this Order or under any other enactment, including any enactment repealed or revoked by this Order, are subject to a suitable system of maintenance and are maintained in an efficient state, in efficient working order and in good repair".

Regardless of any legislation it makes good business sense to ensure that the Fire and Rescue Service can access a building as quickly as possible in the event of a fire.

Generally speaking fire fighting is carried out within buildings. The factors listed below need to be considered so as to assist the Fire and Rescue Service in their task:

- Vehicle access for fire appliances.
- Access for fire-fighting personnel.
- Provision of fire mains in tall buildings.
- Venting for heat and smoke from basement fires.
- Fire behaviour of insulating core panels.

Requirements for vehicle & building access

Vehicle access

The Fire and Rescue Service need to get their appliances as close as possible to buildings to prevent time being wasted with running out unnecessary hose. Building Regulations lay down the minimum access requirements for pumping appliances and high reach appliances. The requirement will vary dependent upon the height, floor area of the building, and if a fire main is fitted. Access will be required for a minimum to 15% of the perimeter or within 45m of every point of the footprint of the building, up to a maximum of 100% of perimeter, dependent upon the above factors. This may differ in practice between one building and another depending on what requirement was imposed on the building when it was built and the level of maintenance.

Access for fire-fighting

In low rise buildings additional access requirements are not normally required as use of the normal means of escape, plus access to the building by the use of ladders, in conjunction with the vehicle access requirements above, are normally sufficient. In other buildings additional facilities, including fire-fighting lifts, fire-fighting stairs and lobbies (usually called a fire-fighting shaft), are needed. The addition of these measures would allow the Fire and Rescue Service to quickly access the fire floor or to access the floor below a fire, from which an operating base can be set up. In general, buildings with floor levels over 18m, or basements more than 10m below a Fire and Rescue Service vehicle access level will need to provide fire-fighting shafts.

Fire mains

A fire main is a vertical pipe installed in the premises with an access point and connections at ground floor level for the fire and rescue service to connect to and pump water up the building. There will then be access points and connections on all floor levels to allow the fire and rescue service to connect to so that they have a supply of water for fire fighting. The fire main will save a large amount of time and effort that would be used by physically running hoses up the stairwells. Any building provided with a fire-fighting shaft will have a fire main fitted. If the building is over 60m above access level then it must have a wet riser installed. Any fire mains fitted are part of the fire strategy for the building and they must be maintained and tested to ensure their safe operation.

Smoke/heat venting of basements

VENTING OF HEAT AND SMOKE

Any basement fire is notoriously difficult to tackle as fire-fighters have to descend down through the heat barrier to make an attack on the fire, or to effect rescues. For this reason basements should have smoke outlets vented to open air so that heat and smoke can be released, therefore making access for fire-fighters possible. This requirement may not be necessary for buildings with small basements.

Another area of buildings that have to be ventilated is the fire fighting shafts (stairwells for access for fire service). These will need to have either fixed venting as seen in the photograph opposite, or mechanical ventilation. This system is not too dissimilar to that utilised in basements.

Insulated core panels

Insulated core panels are a type of building board which is commonly used in construction. It consists of a front and back sheet which is often made of light metal alloy, with a heat insulating filler. The filler can be many materials including: rockwool fibre, polystyrene foam or polyurethane foam.

Figure 4-38: Ventilation to fire fighting shaft. *Source: FST.*

If insulated core panels are installed in a building, a risk assessment should have been carried out at design stage to identify any weaknesses that may be caused to the building structure in the event of a fire affecting the core panels. The possibility of rapid fire spread or structural collapse is a known factor with buildings with a high concentration of certain types of insulated core panels. In such buildings the Fire and Rescue Service may make a dynamic risk assessment not to fight the fire from inside the building but to do so from outside.

General points on access and facilities

Care needs to be taken to ensure that alterations, both internal and external, do not jeopardize these requirements. For example it may only take the erection of a fence, landscaping of gardens or amended parking arrangements to severely affect the access requirement for fire appliances.

It would also be prudent to think of other aspects of Fire and Rescue Service access, for example:

- How accessible is a keyholder for the building out of work hours?
- Does the fire alarm system have a mimic panel externally so that the Fire and Rescue Service can see where the problem is and investigate further without breaking in to the building?
- Are plans of the building easily accessible at the time of the fire, to enable the Fire and Rescue Service to make the best judgment for access and fire fighting arrangements?

Safety of people in the event of a fire

Overall Aims

On completion of this Element, candidates will understand:

- devising procedures in the event of a fire.
- how people perceive and react to fire danger.
- the measures needed to overcome behavioural problems and to ensure safe evacuation of people in the event of fire.
- assisting disabled people to escape.

Content

Specific Intended Learning Outcomes

The intended learning outcomes of this Element are that candidates will be able to:

5.1   advise on the development and maintenance of a fire evacuation procedure

Relevant Statutory Provisions

The Management of Health and Safety at Work Regulations (MHSWR) 1999

The Regulatory Reform (Fire Safety) Order (RRFSO) 2005

The Disability Discrimination Act (DDA) 1995

5.1 - Emergency evacuation procedures

Evacuation procedures and drills, alarm evacuation and roll call

Consideration needs to be given to all who may enter the premises. This should not be restricted to employees only, but should also include any foreseeable visitors to ensure there is a safe means of escape for everyone.

The Discrimination Disabled Act (DDA) 1995 places a duty on persons in control of premises to consider, and where reasonable, to improve access so as not to unreasonably restrict access to those with disability. As a consequence of this legislation people with a variety of disability impairments may now be found throughout buildings on different floor levels. The fire risk assessment should identify when and where such individuals are likely to be present when the building or premises are in use. Consideration should be given to the effectiveness of conventional audible fire alarms to alert anyone with special needs, for example, impaired hearing, to the need to evacuate. Arrangements may need to be in place to assist others, for example, those that require the use of a wheel chair or those with sight impairment.

There is a misconception in some companies that it is satisfactory to evacuate the person(s) with the disability to a safe refuge e.g. fire protected stairwell and to hold them there awaiting the arrival and assistance of the Fire and Rescue Service. The primary responsibility for evacuation lies with the person in charge of the premises and not the Fire and Rescue Service. Evacuation plans will need to consider, as a minimum, the most common disabilities such as mobility, vision and hearing impairment, together with any others specific to the premises

The dangers which may threaten persons in the event of a fire depend on many different factors. Consequently, it is not possible to construct a model procedure for action in the event of fire in all premises. If the fundamental principles are understood it is possible to adapt them to the circumstances of each case. Those who carry out fire risk assessments should be trained in the fundamental principles of fire prevention, detection and evacuation to enable them to determine suitable risk controls, including an effective evacuation procedure.

EVACUATION PROCEDURE

To avoid delay in evacuating the premises when the fire alarm is sounded there should be a pre-arranged plan supported by evacuation procedures, enabling persons to leave safely and quickly.

The essential factor is that all workers and visitors should be familiar with the escape route to be used in the event of fire and an alternative route in case the main escape route be impassable. Therefore all visitors, contractors and employees allowed access to a building should be given instructions about their action on hearing the fire alarm. This would usually be by means of a visitors badge on entering or for employees and contractors as a part of their induction/familiarisation training. The instruction should state which route is to be followed, which alternative route is to be used if the first is impassable and where to assemble for a roll call on reaching open air.

Buildings with public access are a little more complicated to deal with as it is very difficult to train or educate the public prior to their entry into the premises. Therefore a fire management and evacuation system needs to be devised that accounts for visitors and public whilst within the premises. It is not possible to rely on an individual member of the public to read information on the reverse of their visitor's badges and reliably act upon it. Most premises with this scenario develop a fire marshal floorsweep system, whereby all areas are checked as part of the evacuation process. In this way anyone who has not reacted to the alarm will be located and can be directed (or assisted) to escape the premises.

The evacuation procedure devised should set out who will do what, while located in which area, when they will do it and how they will do it. Evacuation procedures vary depending on the type of building, use that is being made of it, those that occupy it, acceptable time allowed for evacuation, and means of escape available.

Evacuation procedures should address the following evacuation options as they apply to the circumstances:

- Single stage (total).
- Horizontal.
- Staff alarm (controlled).
- Two stage.
- Phased.

Evacuation methods

Single stage evacuation (total)

Single stage evacuation involves the immediate evacuation of all the occupants of a building on the sounding of the alarm.

Horizontal evacuation

Certain types of premises may not be suitable for single stage (total) evacuation due to the type of people within the building, for example, in residential care homes or hospitals. In this type of premises it is normal to only evacuate residents from the fire affected area, in the initial stages. Residents should be evacuated to a place of 'comparative' safety. Normally this would mean that there were two fire doors separating the residents from the fire. The place of 'comparative' safety must have another direction of escape that is away from the fire area.

It is paramount that the building structure is maintained so that fire or smoke cannot bypass the fire protection and jeopardize the safety of the people in the place of 'comparative' safety. If the fire threat increases and threatens the place of 'comparative' safety, there must be a method of evacuating residents further away. This is normally done by moving them horizontally away from the fire, behind additional fire resistance. However, it may involve vertical evacuation and if necessary total evacuation.

Staff alarm controlled evacuation

In some premises, an immediate single stage (total) evacuation may not be desirable, for example, in night clubs, shops, theatres, cinemas. A controlled evacuation by the staff may be preferred, in order to prevent distress and panic to the occupants. If such a system is used, the alarm must be restricted to the staff, by the use of coded announcements or direct contact by radio communication, and only used where there are sufficient members of staff present fully trained in the action to take in the event of fire.

Two stage

In a two stage system the first stage alarm signal acts as an alert, to enable the shut-down of operations and the evacuation of the disabled (where appropriate) to commence. This allows a short period of time for the incident to be investigated and if a fire warranting evacuation is confirmed the second stage of the alarm is sounded. If investigation determines a false alarm or the fire has been extinguished the first stage alarm can cease. It is important that the alarms for the first and second stage are notably different and a common, simple way is that the alarm system sounds an intermittent signal for the first stage and a continuous signal for the second. This type of evacuation is useful in situations where it is undesirable for the occupants of the building to evacuate immediately, particularly where work processes may cause false alarms or small occasional fires that are extinguished by local fire-fighting equipment.

Phased evacuation

Phased evacuation systems are becoming more common as buildings become taller and occupancy levels increase. In this type of evacuation the occupants of the floor on which a fire occurs together with the occupants of the floor directly above are evacuated immediately a fire is detected. The remainder of the occupants are alerted and remain in the building unless the fire is such that they need to evacuate. Depending on how the fire develops and spreads, further evacuations are carried out two floors at a time under the control of fire marshals or the Fire and Rescue Service.

Good communications are essential if an orderly evacuation is to be undertaken. If a phased evacuation system is in operation, fire marshals will need to be able to communicate with the senior fire marshal in a 'Control Centre' by fire telephones, which should be installed on each floor of the building.

Buildings in which phased evacuation is proposed should incorporate the highest standards of fire protection, and staff should undergo comprehensive training. In particular, each floor should be constructed as a compartment floor, automatic fire detection and alarm system should be installed, and smoke control measures are recommended. If the building is over 28 metres high, an automatic sprinkler installation should be considered.

Allocation of Responsibility

Allocation of specific responsibilities will vary according to circumstances. The most important requirements are that there should be no confusion as to who is responsible for each of the various measures that may need to be taken, and that decisions are not delayed unnecessarily because one or more of those responsible are not immediately available.

The ideal is that there should be a designated person in each department, or on each floor, responsible for taking immediate charge at the scene of the outbreak and for taking decisions which may be required pending the arrival of a more senior member of the managerial staff, the premises fire officer, or the Fire and Rescue Service.

It should be clearly understood that if the designated person is not available then the deputy, or next senior person, takes charge.

The tasks of the designated people in the departments are to ensure the safety of the occupants, and if the fire is in their department, to attempt to contain it - if trained to do so. When in doubt, personal safety and evacuation is the more important consideration over tackling the fire.

Fire Instruction Notices

At conspicuous positions in all parts of the building and adjacent to all fire alarm actuating (break glass) points, printed notices should be exhibited stating, in concise terms, the essentials of the action to be taken upon discovering a fire and on hearing the fire alarm. Individuals should not be encouraged to tackle fires unless they have been trained in assessment of the risk and the use and limitations of any fire-fighting equipment.

Fire Action

In the event of a fire, action upon discovery needs to be immediate and a simple fire action plan should be put into effect. This is reflected in the fire instruction notices. A good plan of action would include the following points:

1   Sound the fire alarm (to warn others).

2   Call the Fire and Rescue Service.

3   Tackle the fire (only if trained, and it is safe to do so).

4   Responsible person for each area checks area clear of personnel.

5   Get out of the building and stay out.

6   Carry out a roll call.

Procedure to be adopted in the event of fire

In drawing up the procedure, care must be taken to ensure that:

- A senior member of staff is responsible for fire safety (which will include fire prevention and action to be taken in the event of a fire).
- All members of staff are given clear and comprehensive instructions.
- Responsibilities in the event of fire are clearly allocated.
- The design of the premises and the particular hazards of the materials and processes used in the premises are taken into account.
- The alarm system is clearly audible and distinguishable throughout the premises.
- The fire 'control centre' - probably the telephone switchboard - will come into operation in the event of fire with standing instructions as to the immediate actions to be taken - this includes summoning the fire brigade and notifying the management.
- The evacuation procedure is efficient and all employees and others are familiar with the main and alternative escape routes from their department.
- In the case of evacuation, provision is made for a search of every part of the premises.
- Practice fire drills are held at least once a year.

THE ROLL CALL SYSTEM

The roll call system is based on checking that everyone in a building has reached a place of ultimate safety. The names of all the building's occupants are recorded on a list to confirm their arrival at a nominated assembly point. In the event of a fire or practice, following the evacuation of the building, a designated person checks that everyone on the list answers the roll call. This information is then passed to a central control point.

Advantages of this system include:
- Specific confirmation that staff are out of the building and safe.
- The emphasis lies in getting people out then checking.

Disadvantages of this system include:
- It is dependent on complete and up-to-date lists of the building's occupants, which are often difficult if not impossible to maintain at any one time.
- It is reactive to evacuation - it doesn't help to get people out.
- A lot of time is spent checking lists before the area can be declared clear.
- It assumes all personnel know where to go if the building is evacuated.
- Nominated people, and their substitutes, must be available to conduct the role call at the time they are needed.

THE FIRE MARSHAL SYSTEM

This system is based on splitting a building into small manageable areas. In the event of a fire, designated people (Fire Marshals) will search their area to ensure that all people leave the building. The fire marshals then direct any people who have not evacuated to the appropriate fire exit and onward to their assembly area. They then report that their area is clear, or otherwise, to an allocated person at the assembly point.

Advantages of this system include:
- It has been shown to be the quickest, most efficient way to evacuate a building.
- It allows the Fire and Rescue Service into the building quickly to rescue people and reduce damage.
- Buildings are split into pre-defined areas for control - no 'grey' areas e.g. rooms used to store cleaning materials.
- It is a pro-active approach - fire marshals identify dangers and problems arising during - not after evacuation.
- The system uses people to evacuate people and by doing so it allows for adverse human behaviour.
- It allows for a controlled search of an area defined by information from the fire marshals, if a search is necessary.

Disadvantages of this system include:
- It may only be in operation during the core working hours for the building.
- The role of fire marshal is normally voluntary - it relies on staff goodwill and their participation.

It is important that all areas of the building are covered by fire marshals, and organisations who choose to adopt this system must ensure that there are sufficient numbers of fire marshals to cover the building at all times, bearing in mind that 'extra' fire marshals will be required to cover absences. Absences could include sickness, holiday leave, attendance at training courses, or people temporarily absent from their normal place of work, for example delivering a report to another part of the building.

The actual system of work operated by the fire marshals will be dictated by the building layout, the work practices and number of staff available to conduct an evacuation. In practice one of the three options outlined below, or an adaptation of these, will be used. In all cases the fire marshals will check areas within their nominated area and instruct staff to evacuate the building by the nearest safe exit. The fire marshals will be the last person to evacuate the area they are checking. It is always advisable to give fire marshals some visible form of identification such as high visibility jackets as well as some form of communication equipment so that they can communicate with the person in charge of the fire panel. In this way the fire marshals located within the area where the fire alarm has been activated can be made aware of the potential for fire by the person in charge at the fire panel and in return the fire marshals can make the person in charge at the fire panel aware of any issues they are confronted by (e.g. mobility impaired awaiting assistance in safe refuge).

Fire marshal - fixed system

In this system people occupy and use the building in such a way that they (or a portion of them) do not normally move around the building during their normal work operations. People who are therefore generally at fixed positions within the workplace can be selected as fire marshals. They can then be given a fixed area of the building as a search area, and when the fire alarm operates they can check their area before leaving the building. Providing that there are sufficient fire marshals at work that all areas can be covered this system is the simplest type of 'floor sweep' method to be used.

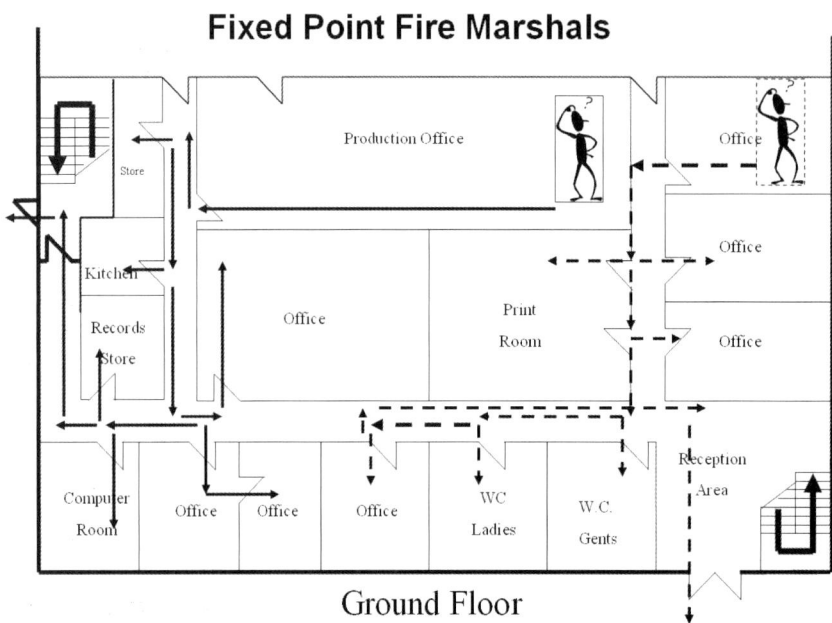

Figure 5-1: Fixed point fire marshals.   *Source: FST.*

Fire marshal - assembly systems

This system is useful if it cannot be guaranteed that people acting as fire marshals are in fixed positions in the building. In the assembly system, on actuation of the fire alarm, the fire marshals make their way to a predetermined point, or points, within the building. From there they can then be dispatched to check a specific area. The end result is the same as in the fixed system as all areas are checked, the difference being that the start points for the fire marshals are changed. This system may be slightly slower than the fixed system as the fire marshals have to assemble and then be dispatched to make their checks. The advantage with this system is that it will work even if the number of available fire marshals varies.

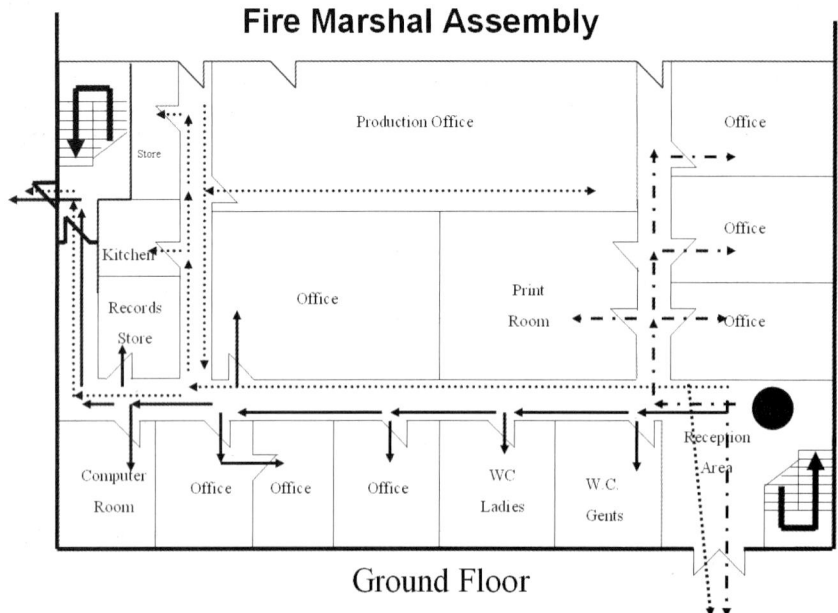

Figure 5-2: Fixed marshal assembly.   *Source: FST.*

Fire marshal - points system

In this system, again the number of available fire marshals can vary and they may be constantly moving around the building. At specific sites around the workplace 'Fire Marshal Points' are located. At each point is a route card showing a specific floor sweep area. On actuation of the fire alarm the fire marshal goes to the nearest fire marshal point to their position at that time. They would then check the area denoted on the available route card. This system can be as quick to operate as a fixed fire marshal system. However, its speed of operation will depend upon the number of available fire marshals. If many of the nominated fire marshals move around the building, particularly on different floors there is a risk that a given area be depleted of fire marshals and another have too many. It may be possible to direct a fire marshal towards a nearby floor but they should not be put at unnecessary risk. If this situation was the case it would be advisable to revise the nominated fire marshals and establish more fire marshals who did not move from a fixed point so often.

To operate this system most effectively a nominated person would need to be controlling the evacuation from a safe point, for example, the fire alarm panel, and they would need to be in communication with the fire marshals. This can be relatively easily achieved by placing portable radio communications equipment at the fire marshal points, so that the co-ordinator can be in constant communication with the fire marshals concerned.

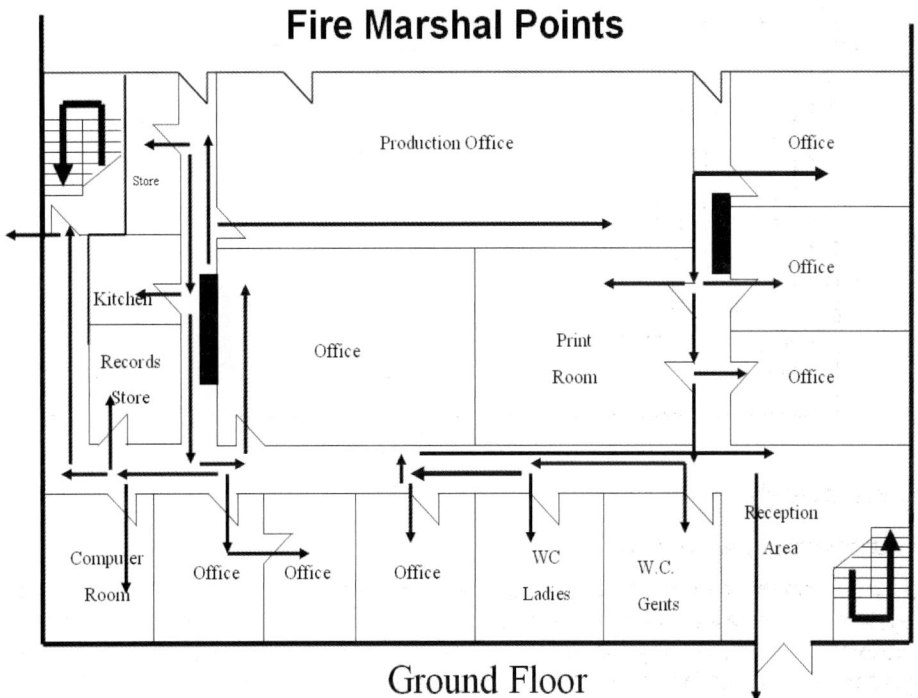

Figure 5-3: Fire marshal points.   *Source: FST.*

In all cases the fire marshals need to report to a 'Co-ordinator' at the assembly point, so that complete evacuation of the building can be confirmed. This information, plus any additional relevant information, should be passed on to the Fire and Rescue Service on their arrival.

In addition to the designated fire marshals, specified people should be chosen to perform any other tasks which may be necessary in the event of fire and should be given explicit instructions stating exactly what they should do. Such tasks may include guiding the Fire and Rescue Service, shutting down conveyors and machinery, and safe guarding dangerous processes.

FIRE DRILLS

To ensure that all employees are familiar with and understand the procedure in the event of fire, repeated practice is desirable. After initial practices to establish the procedure, practice drills should be held at least once a year. Where there is danger of rapid fire spread, more frequent drills are advisable.

An additional method of familiarising employees with escape routes, which will not interrupt normal work, is to instruct them occasionally to leave by their main or alternative escape routes at the end of a working day. This will need to be undertaken with the assistance of the security staff, who may have to guard some vulnerable fire exits while they are insecure, and who will have to check the security of the building at the conclusion of the exercise.

ALARM EVACUATION

To avoid delay in evacuating the premises when the fire alarm is sounded, there should be a pre-arranged plan enabling persons to leave safely and quickly.

People with sensory impairment

The principal sensory impairments that will present issues with alarms and evacuation will be:

Hearing Impairment — Evacuation may be assisted by the issue of personal trembler alarms, flashing lights and 'buddy' (work companion assistance) systems or the equivalent for visitors to the building.

Vision Impairment — Evacuation may be assisted by use of tactile way-finding and exit signs and 'buddy' systems or the equivalent for visitors to the building.

Hearing and vision impairments — If an individual has both a hearing and vision impairment, a combination of both the approaches discussed above will usually be necessary to ensure safe evacuation.

Procedures to evacuate disabled people

Mobility Impairment — This may vary from someone who is just slower than everyone else in escaping, to a person in a wheelchair. Solutions may vary from assisting the person out of the building immediately after the initial rush of occupants, through the use of evacuation chairs or other escape systems. In the case of a person in a wheelchair being required to evacuate vertically there may be significant risk factors involved in moving them. Depending on the risks to the individual it might be appropriate to utilise a safe refuge. An appointed person would need to remain with them to maintain communication between the refuge and the co-ordinator at the fire panel. This would then enable an informed judgement to be made on the need to evacuate the individual further or not.

See also - use of evacuation lifts and refuges - Element 4.

Figure 5-4: Refuge point. *Source: FST.*

Figure 5-5: Access equipment on means of escape. *Source: FST.*

Safe refuges — A safe refuge is an area of the building that people who may need assistance to evacuate can go to. The area would generally be a fire protected area with access to an evacuation lift or a fire protected stairwell. The refuge may in fact be an area within the normal escape stairwell, but sited so that it does not inhibit the escape route for others. The refuge should be treated as a 'temporary safe haven', where people can initially be evacuated to whilst investigations decide if it is necessary to evacuate the individuals further. In all cases the onus on evacuating people belongs to the responsible person and they should have a safe system of work, equipment, resources and staff to carry out any evacuation if needed.

Actions required when evacuating members of the public

There will generally be greater problems with evacuating the public than evacuating employees. There will be a variety of issues that influence this, including factors such as:

- Disbelief that a fire is occurring.
- Lack of awareness of the dangers.
- Unwillingness to stop what they are doing at the time - shopping, eating, exercising, carrying out an 'important task' that took them a long time to get to that point.
- Unwillingness to break their pattern or 'script' - their normal pattern of behaviour under non fire conditions.
- Lack of knowledge of evacuation routes and evacuation systems - leading to some resistance to go in an unfamiliar direction where they might need to be directed.

It is not possible to make individual escape plans for all people who enter a public building, whereas with staff it is possible to pre-plan with the individual to develop a Personal Emergency Evacuation Plan (PEEP). All that can be done for visitors and the general public is a generic assessment of the type of issues that they may be confronted with together with a set of generic evacuation plans for them to use. The use of competent fire marshals is an essential component of a well thought out public evacuation plan.

5.2 - Perception and behaviour of people in the event of a fire

HUMAN BEHAVIOUR IN FIRE

To many people, involvement in a fire may be their first experience of immediate personal danger. It is very difficult to predict how a specific person or groups of people will react to this danger, due to the individuality of people. We can therefore only look at this subject in general terms, perhaps looking at some factors that may affect people's behaviour at the different stages of a fire.

Most people will display fear in some way, but will eventually evacuate a building pending the arrival of the Fire and Rescue Services. Others, however, may display a neurotic fear which is out of proportion to the danger. This neurotic fear may lead to panic.

Before looking at these factors and stages of behaviour we need to first consider the people involved and the training that they have received. Research has shown that the more frequently people receive training in what to do when a fire occurs, the more likely they are, in the event of an actual event, to act rationally, to help others, and to remember to raise the alarm. In addition, the more familiar a person is with the layout of the building and the escape routes, the more likely that person will attempt to move through smoke, in the knowledge that a place of safety is only a short distance away. Training may also help to subdue the individual's desire for self-preservation at the expense of others.

When confronted by a fire a well disciplined and well trained work force should automatically raise the alarm, ensure the Fire and Rescue Service is called and evacuate the building. However, this not a guarantee as human behaviour in a fire is a varied and complex reaction.

In order to begin to understand human behaviour in the face of fire we need to consider the factors that affect behaviour at the different stages of a fire.

It has been found on frequent occasions that people do not treat the dangers of fire with the respect that it deserves. In some cases the initial reaction to a fire is one of disbelief, complacency, interest in watching the outcome, or other forms of irrational behaviour. This creates problems in evacuation as a group of people have to be sufficiently motivated to begin to escape. Once people have started to evacuate it is generally found that at this stage others will follow on, displaying what some call "The Sheep Syndrome" which is the peer pressure which drives people to act as a group.

The initial reaction stage in people's behaviour will continue until the individuals perceive that they are in danger, and it is at this stage that they will look to evacuate. The problem, however, caused by this fact is the difference in people's perception of when they are at risk. All people have a personal psychological safe/unsafe boundary line set at a distance from them. It is only when the danger, e.g. fire or smoke, encroach upon their personal boundary line that they perceive they are at risk. The distance at which this line is set varies from person to person, and so therefore, does their response times.

If people do not respond quickly enough to the dangers of fire, they can be overtaken by the speed of fire. People are then engulfed in smoke or endangered by flames before they have begun to evacuate.

It is at this point that the need for self-preservation, irrational behaviour or panic can occur.

However, this process is further complicated by people's reactions at the stage that a fire is in progress, especially if someone is attempting to extinguish the fire. It has been found that persons can then become interested, almost fascinated, in watching to see the result. For example, in the "Stardust Nightclub Fire" in Dublin, people stopped dancing and walked away from the exits to watch an attendant attempting to extinguish the fire.

Two significant factors that will affect individual behaviour in the event of a fire are the effects of smoke and personal relationships. In general it has been found that some people are unwilling to travel through a smoke filled area even if they know that there is an exit within reach. The maximum distance that persons will travel through smoke is normally 6 metres. This factor can be affected by the training that persons have undergone and their knowledge of the building layout.

People's protective nature towards family, friends and especially children can also play a part. The "Summerland" fire in the Isle of Man highlighted this fact. Parents, who were separated from their children, ran in completely the opposite direction from the escape route in order to locate their children, thus placing themselves at increased risk. Where crèche facilities are provided in shopping malls for the young children of shoppers, great emphasis should be made by the crèche staff to parents, at the time of placing the child with the carers, to ensure that the parents know that their child will be taken to a defined point of safety at the first indication of a fire alarm being raised. The parents then know where to go to meet them with certainty and should see no reason not to obey an instruction to evacuate.

As can be seen, human behaviour in fires is a very complex matter, but it is an area that must be considered when planning fire safety in buildings.

Centuries of experience have provided some knowledge of how to control the physical effects of fire through legislation, development of good means of escape, including structural design of buildings and other engineering aspects to minimise fire/smoke spread.

Compliance with the fire safety requirements of building regulations, standards and codes reduce the fire risk in a building and provide adequate means for safe evacuation. The success of these measures, however, depends to a great extent on the behaviour of the building's occupants at the time of the fire. Buildings contain a variety of people, some who will be able to escape in most circumstances, others who would have extreme difficulty and those who still do not attempt to escape (among whom some will take the risk of fighting the fire). These facets of behaviour are influenced by psychological and physiological factors and previous involvement in fire incidents. They can also be influenced by the severity of the threat posed by fire, the building design and the fire protection devices installed.

Main conclusions

(a) In a fire the behaviour of individuals can be different, accordingly to the role in which individuals see themselves and whether that role is one acting in a strong hierarchy. For example, in hotels staff behave differently from guests and in hospitals staff may continue with their health care duties to patients.

(b) Building occupants do not use escape routes that are unfamiliar to them. This implies that purpose-built fire escapes may be less effective unless management's role is clearly defined so that the staff guide people to the evacuation routes. Generally it is now thought that a fire exit sign is less important than the regularity with which a route is used. Too much emphasis may be placed on travel distance and time of evacuation. Alternative routes which are part of the normal circulation routes in the building are far more effective in an emergency. Familiarity may also decrease the "time to move" part of the equation of evacuation time thus speeding evacuation even though it may be further to travel than an unfamiliar route.

(c) Fire extinguishers are rarely used and are not very effective without training.

(d) The earliest cues to fire are generally strange noises like breaking glass and extra activity by others, rather than flame or smoke.

(e) Early behaviour is characterised by uncertainty, misinterpretation, indecisiveness and seeking additional information for confirmation - "the gathering phase". Such delay can be dangerous, as actions taken at the early stages of a fire have the most decisive effect on the eventual outcome.

(f) The response to fire alarm bells and sounders tends to be less than optimum. There is usually scepticism as to whether the noise indicated a fire alarm and if so if the alarm is merely a test or drill.

(g) In the stress of a fire people often act inappropriately and rarely panic or behave irrationally. Such behaviour, to a large extent, is due to the fact that information initially available to people regarding the possible existence of fire, its size and location is often ambiguous and inadequate.

(h) The concept of 'panic' is possibly unjustified as this reaction often results when there is a serious delay in being made aware of a fire threat. As intimated previously, people are far more familiar with the normal routes into and out of the building than an alternative emergency route. If they therefore receive a fire warning too late to reach a place of safety easily via the routes they are aware of, it should not be surprising that they will act in a stressful and sometimes what appears to be an irrational manner.

The following general factors will affect an individual's escape behaviour:

- Importance of building dimensions.
- Own role within the building.
- Location of individual.
- Knowing there is a fire.
- Awareness of the location of the fire in relation to the individual.
- Familiarity with the route to safety.
- Reaction to smoke development.

Principles of sensory perception

EARLY RECOGNITION BY THE SENSES

Fire engineers must ensure that building occupants become aware of potentially dangerous fire situations in time to take appropriate action. One way of doing this is by providing clear, consistent and easily recognisable cues. Consequently, fire engineers must have a sophisticated understanding of how people perceive fire-related stimuli.

Such stimuli include obvious signs of fire such as:

- Visible smoke or a smoke odour.
- Radiant and convective heat from a proximate fire.
- Arcing, sparks or visible flame.

Of course, many of these cues pose the danger of physical harm. Exposure to excessive concentrations of carbon monoxide or other toxic compounds in smoke can cause hypoxia. High radiant or convective heat exposures can result in serious burns. However, these cues generally require little interpretation before individuals appreciate the danger present in a situation.

The following signs of fire require individuals to identify the source of the information or place it in some sort of situational context to appreciate its relevance.

- Bells, horns or other audible warning signals.
- Loud indistinct or otherwise unrecognisable noises.
- Flashing lights or a sudden flash of light.
- A power outage or a sudden loss of telephone communication.
- A sudden movement of other building occupants towards exits.

Information that appears incongruous or irrelevant will often provoke a human response, but this response may not be constructive in dangerous or threatening situations.

RECOGNITION OF FIRE THREAT

Among the most common ways of ensuring that people perceive fire danger in time to react properly is through the installation of mechanical or electrical systems to detect fire cues (signatures) and translate them into signals that will be interpreted as a fire warning. Fire detection and alarm systems can expedite the detection of fire cues and shorten the time between detection and response by translating early fire cues into a warning. Under ideal conditions, such systems can also provide an indication of the location, intensity and nature of the fire threat or suggest an appropriate response. However, proper interpretation and action on fire warning signals or instructions requires training. People can be conditioned to disregard signals through frequent exposure to false or nuisance alarms.

Perception v. reality

The perception of people's reactions in a fire situation is that they will panic. Studies of people's reactions and history itself has proven that this is not necessarily the case. People confronted with information about a fire situation will usually do any one of the following things:

- Take no action - ignore or fail to recognise cues.
- Wait for additional information - recognise cue.
- Investigate or explore the situation.
- Warn others.
- Instruct others.
- Withdraw - flight.
- Evacuate - escape.
- Fight fire.
- Freeze - fail to respond.

As stated above, the general view is that people will panic when confronted by a fire situation. Panic is defined as a sudden overwhelming feeling of terror or anxiety, especially one affecting a whole group of people. At its most extreme, it can lead people to do things of which they have no later recollection. In a fire situation, panic is most likely to result from the belief that the individual is trapped, particularly if all the people around him/her are in the same position. This fear is more likely to be intensified if an escape route can be seen but is not immediately available, possibly due to the number of persons trying to escape through it causing a blockage.

Experience has shown that in general people do not panic specifically in the early stages of a fire.

Response to different forms of audible and visual warnings

For a warning device to be successful it has to be noticed, recognised and acted upon by the people at risk. There are two basic types of warning device - audible and visual.

Audible warnings

There are three ways in which audible warnings can be used:

Bells/sirens	The traditional fire alarm bell or siren has to be loud enough to be audible throughout the premises, but not excessively loud so that they contravene the upper noise limit set by the noise legislation. They need to be discernible from other alarms and sounds that are produced within the building.
Verbal messages	Fire warning systems in public buildings can be linked to a public address system so that a pre-recorded message can be played. This system gives people far more information than the traditional sounder or bell and as a result should speed up personal reactions and the evacuation process.
Siren and verbal	In general the modern fire alarm system in public buildings will in fact be a mixture of both of the above systems. A bell/siren will sound first to get people's attention so that they are then more receptive to the recorded message.

Visual warnings

There are generally two ways that visual warnings are used:

Strobe lights	Flashing strobe lights may be used as an alternative, or as a backup, to the audible warning devices. This type of system can be very useful for giving warning to people with hearing impairments or in areas where the background noise levels are high.
Information boards	In modern systems, particularly in public buildings, the fire warning system can be connected to large electronic message boards. These can then be used to give directions to people in order to speed up the decision making process and enable evacuation to be quicker.

Negative aspects of warnings

Bells/sirens	This type of system relies upon an audible noise to warn of danger and its effectiveness is dependent upon whether not it gets people's attention. The major weakness is that this type of system relies on people to react to the warning and evacuate themselves. To work efficiently it normally needs to be reinforced by the use of fire marshals.
Verbal messages	On their own this system has a weakness as people would treat it as a normal announcement and may not listen to or take heed of it properly.
Siren and Verbal	The combination system appears to be a very successful system at intercepting human behaviour. The combination of the audible noise followed by the taped message tends to get attention better so that people listen to the message. The main limitation, in a multicultural society, is the potential language barrier.
Strobe lights	The sighting of the flashing lamp/strobe/beacon is paramount, as is its light intensity. The effectiveness of strobe lights will depend upon the tasks people are doing at the time. For example, if an individual is deeply involved in watching a computer screen they may not notice a light that is flashing in their peripheral vision and it may not get their attention. A quicker response may be achieved if their computer screen was linked to the alarm system causing the screen on the computer to flash at the same time. Evacuation response is similar to that of audible alarms.
Information boards	Similar to a pre-recorded message, the information board will give more comprehensive information so that people can make better and quicker choices for evacuation. Again a possible limitation would be language and / or literacy barriers.

The decision making process during fire emergencies

A fire situation or an evacuation which gives a time limitation will create stress to those individuals affected. It is therefore worth considering how stress affects people in a fire situation and to use this understanding at the design stage of evacuation systems. In a fire the range of vision will often be reduced due to smoke and fume emission. This means that reliance on only visual symbols will be insufficient in an emergency situation. In every emergency situation it is crucial that information is precise and easy to comprehend. Because of the effect of cognitive tunnelling (tunnel vision), it can be difficult to reach out for the vital information. Research suggests that evacuation orders given by voice or text should tell people *what to do*, not what they should not do. It is considered easier to understand orders about which actions to take than to understand what actions one should avoid, and this accounts for signposting as well. The order of instructions should also be chronological - "first do A, then do B" is easier to understand than "before you do B you must do A". Generally, symbols and pictographs are easier to understand than text, but even better is a combination of picture, text and speech. As for speech, words used regularly are easier to understand than rarely used terms. *(Based on Stene, Jenssen, Bjørkli & Bertelsen, 2003)*

Spatial orientation and way-finding in large and complex locations

Due to the factors highlighted above, especially the tunnel vision effect on people's eyesight range, people exiting large public buildings will tend to be fixed on the exit route path they have chosen. They will not tend to scan around the area visually to look for other exit routes or signage and will often completely miss exits that are near at hand, to utilise the route they thought of first. It is only by thinking of the way people behave that we can properly design fire safety into buildings.

For example, the entrances / exits into a departmental store may be oversubscribed compared to the number and size of exits theoretically needed. By taking this approach we account for the human behavioural trait of people exiting out of buildings by the way that they entered.

If we need to 'steer' people then this can be re-enforced by the use of signage and wayfinding systems. For example, in a cinema the exit routes are permanently lit with low level lights. As we sit in the cinema in the dark, this exit route system is then imprinted into our subconscious memory without us realising it. Should there be a need to evacuate then there is a greater chance that we will use the exits located within the screen room where we are sitting rather than exiting out into the concourse in order to get back to the main entrance.

Patterns of exit choice in fire emergencies

The majority of people do not generally walk around buildings looking at the emergency exits. If therefore they have to evacuate from an unfamiliar building they will tend to try to reverse their steps and exit via the route (known way) that they came in by, even if this takes them past other exits. Those who are more familiar with a building will tend to use the exit routes that they are familiar with which may not be the way they originally entered. These statements highlight the reasons why practice fire drills are vitally important and also why we should simulate a blocked exit route during a fire drill evacuation. It is only by forcing people to use their alternative exit routes that these routes will become a little more familiar and hopefully people will then be more liable to use these routes in the event of a real evacuation. We cannot undertake this procedure every time a visitor enters the building so we must re-enforce this message to them as part of the visitor induction process. Alternatively of course we can control the visitor at all times with our staff who will be well versed in the correct evacuation processes to take.

THE IMPLICATIONS OF EXIT CHOICE BEHAVIOUR IN DESIGNING FOR FIRE SAFETY

When designing buildings therefore it is questionable to just consider the number of people in the building and assume an even split of people to exits. We know for example that people will try to get out by the way that they got in, so by design we should oversubscribe the exit availability at this point. It is essential in the design of large buildings to consider the likely flow patterns at time of emergency. If the building is one such as a leisure centre with multiple function use this should be factored into the considerations and the various scenarios considered.

Behaviour of individuals responsible for others during a fire

PARENTS AND ELDER SIBLINGS

History has proven on a number of occasions that when evacuating from a fire situation parents or elder siblings will generally not evacuate a building willingly if they believe that their children or younger siblings are still inside. If for example parents have left their child at a crèche whilst they go shopping, they will want to return to the crèche to collect their child before evacuating. It is imperative that this side of human nature is accounted for and is built into the fire warning system and the management actions for control of the fire evacuation situations. For example, where crèche facilities are provided in shopping malls for the young children of shoppers, systems should be considered that ensure the children are evacuated at the earliest time that a fire incident is discovered. This system in turn must be communicated with great emphasis by the crèche staff to parents at the time of placing the child in the crèche with the goal that parents will evacuate by the nearest exit, confident that the child is in a known safe location.

NURSES

Similarly nurses will often be very reluctant to leave patients who are under their care to fend for themselves. As an example, it is considered potentially dangerous to enter a room that's on fire. However, a standing instruction to a nurse in a care home 'not to enter a bedroom that is on fire', when the resident is still in the room is likely to bring enormous psychological and emotional effects upon the nurse and the instruction may not be followed.

TEACHERS

Similar issues arise with teachers, especially where there are young dependent children involved.

As you can begin to imagine you can apply these issues to many roles within the community. It is vital that as part of the fire risk assessment process we take into consideration the occupants of the premises and their expected actions and reactions. It is only in this way that we can truly begin to design premises and fire management systems effectively.

Crowd movement as individuals and in groups

Some common assumptions extracted from an article by Jonathon Sime:

(a) People's safety cannot be guaranteed since in certain circumstances they panic leading to inappropriate escape behaviour - in reality deaths in large scale fires attributed to panic are far more likely to have been caused by delays in people receiving information about a fire.

(b) Individuals start to move as soon as they hear an alarm - in reality fire alarm sirens cannot always be relied upon to prompt people to immediately move to a place of safety.

(c) The time people take to evacuate a floor is primarily dependent on the time it takes to physically move to and through an exit - in reality the start up time is just as important as the time it takes to physically reach an exit along with other behaviour on the way to the exit.

(d) Movement in fires is characterised by the aim of escaping - in reality much of the movement in the early stages of fires is characterised by investigation, not escape.

(e) People are most likely to move towards the exit they are nearest to - in reality as long as an exit is not seriously obstructed, people have a tendency to move in a familiar direction, even if further away, rather than to use a conventional unfamiliar fire escape route.

(f) People move independently of each other (unless in a dense crowd) - in reality individuals often move towards and with group members and maintain proximity as far as possible with individuals with whom they have emotional ties, i.e. take refuge with others.

(g)   Fire exit signs help to ensure that people find a route to safety - in reality fire exit signs are not always recalled or noticed and may not overcome difficulties in orientation and way finding imposed on escapees by the architectural layout and design of the escape route. In one shopping mall, tests showed that people instinctively moved in one direction and looked around for exits, they often missed signs next to them if they happened to be looking the other way and some even went upstairs for exits they could see on upper levels when they were available at ground level had they simply stood still and looked all around them.

(h)   People are unlikely to use a smoke filled escape route - in reality people are often prepared to try to move through smoke. Visibility must be less than only 2 metres to deter movement in smoky conditions.

(i)   All the people present are equally capable of physically moving to an exit - in reality people's ability to move towards exits may vary considerably, for example, a young fit adult compared with a person who is elderly or has a disability.

Figure 5-6: Article extract.   *Source: Jonathon Sime.*

HOW CROWD FLOW CAN CAUSE DANGER AND PROHIBIT SAFE ESCAPE

If there are large numbers of people involved in an evacuation we must consider the dangers involved. If a crowd of people is flowing in one direction it is virtually impossible to go against the flow. Unfortunately people have been crushed, injured and killed during evacuations involving such circumstances. If an individual should be unfortunate enough to fall or be knocked to the ground others may trample on them, not by choice, but as they themselves cannot resist the flow of people. The pressures and forces involved with such movement are immense and these alone can be sufficient to cause crush injuries or death. Tragically many disasters of various types have proven this point in the past.

A recent event that highlighted this issue did not even occur during an emergency situation. It was the tragedy at the football match between Liverpool FC and Nottingham Forest in an FA Cup semi final at a neutral ground in Sheffield. A crush grew as turnstiles struggled to cope with thousands of fans who had arrived late due to traffic delays and were trying to gain access to the ground. Various actions were taken to try to relieve the pressure, at one stage a gate was opened and in five minutes around 2,000 fans moved through this gate at a brisk walk, most heading straight down the tunnel which led at a gradient of 1 in 6 to a tightly fenced pen of terracing in the lower West Stand. This was an area already completely full. The gradient and momentum created a domino effect as thousands lost their footing and became unable to control momentum or direction. Unchecked, more and more behind were swept along with the crush. *Fans spoke of being swept through, feet completely off the ground.* Pressure became unbearable People became crushed against crash barriers and high metal fences at the front of the pen. Tragically 96 people died as a result of their injuries.

Similar situations have arisen at fire incidents. Survivors of the Bradford City Football Club fire reported being swept along the rear passageway at the rear of the stand, unable to resist the pressure of the flow of people.

Modification of crowd flow by physical design and messages

Care must be taken therefore that we design buildings to try to prevent injury due to crowd movement in a fire situation. The first consideration would be the provision of directions, signage etc to encourage people to use as many exit routes as possible, rather than large amounts of people trying to get out the way they came in - this would be particularly applicable to a night club or cinema where people generally come through a single entrance. We should also consider each individual route to see how we can aid the flow of people. This can work in lots of different ways. As an example, if people are exiting down the staircase in a multi floor building and the stair flights are wide, people will tend to stay at the sides near the handrail. This would therefore mean that we would not be utilising the full width of the escape route and as such would be causing a detrimental effect on the evacuation process. If however, we split the large open stairwell down into narrower parts by the use of a handrail up the middle of the stairs people will tend to use all 3 handrails (the middle one from both sides) so that we in effect double the number of people who use the stairs at the same time.

Fire risk assessment

Overall Aims

On completion of this Element, candidates will understand:

- the process of fire risk assessment.
- fire risk assessment recording and reviewing procedures.
- the potential for pollution arising from fires.
- measures to prevent and reduce fire pollution.

Content

Specific Intended Learning Outcomes

The intended learning outcomes of this Element are that candidates will be able to:

6.1   use a simple fire risk assessment technique to determine risk levels and to assess the adequacy of controls

6.2   advise on improvements to control measures to reduce the risk of fire

6.3   advise on steps to minimise the environmental impact of fire and fire fighting operations

Relevant Statutory Provisions

The Regulatory Reform (Fire Safety) Order (RRFSO) 2005

The Water Resources Act (WRA) 1991

6.1 - Fire risk assessment

A fire risk assessment is an organised and methodical examination of premises along with the activities carried on to establish the source and likelihood that a fire could start and cause harm to those in/or around the premises.

Definition of hazard and risk

HAZARD

"something that has the potential to cause harm (loss)"

"the potential to cause harm, including ill-health and injury, damage to property, plant, products or the environment, production losses or increased liabilities"

Figure 6-1: Definition of hazard. *Source: Successful Health and Safety Management, HSG65, HSE.*

In this case we are particularly interested in hazards that have potential to cause harm through fire or explosion.

Figure 6-2: Hazard - 4 way extension lead. *Source: FST.* Figure 6-3: Extension into extension - increased risk. *Source: FST.*

RISK

"the likelihood of a given loss occurring in defined circumstances"

"the likelihood that a specified undesired event will occur due to the realisation of a hazard by, or during, work activities or by the products and services created by work activities"

Figure 6-4: Definition of risk. *Source: Successful Health and Safety Management, HSG65, HSE.*

For example, the risk from an overloaded electrical circuit (when a multiple socket device is used to power other devices that cumulatively exceed the rated current capacity of the multiple socket device) is the likelihood that it will lead to a fire. This will depend upon:

a)   The circumstances presented by the hazard.

b)   How it is controlled.

c)   Who is exposed to the hazard.

In the example of an overloaded circuit, major determining factors are, how long the overload exists, any over current devices that may limit its effect, also its relative location to sources of fuel and people.

Outline of the objectives of risk assessment

A fire risk assessment has three objectives:

1)   To identify all factors which may cause harm to people, property and/or the environment either during or as a result of a fire.

2)   To consider the likelihood or chance of that harm actually happening, and the possible consequences that could come from it.

3)   To enable the 'Responsible Person' to plan, implement and monitor the preventive and protective measures to ensure that the risks are controlled as low as reasonably possible at all times.

When undertaking a fire risk assessment there are six steps that need to be taken:-

Step 1   Identify potential fire hazards.

Step 2   Decide who may be in danger, and note their locations.

Step 3   Evaluate the risks and adequacy of current control measures.

Step 4   Record significant findings and action necessary.

Step 5   Prepare action plans to carry out any necessary improvement measures.

Step 6   Keep assessment under review.

The purpose of the assessments is to identify where fire may start, the people who would be put at risk, and to reduce the risks where possible. This type of assessment supplements those already carried out under existing legislation for general health and safety reasons.

The following all need to be considered when carrying out fire risk assessments:

■   Identify ignition sources - consider, where possible, any actions to reduce the risk of causing a fire.

■   Identify combustible materials in the workplace - consider appropriateness of storage, arrangements to reduce the risk of contact with ignition sources.

■   Identify those substances which are oxidising agents and/or those that present a significant fire risk - consider appropriateness of storage, away from sources of ignition.

■   Identify those people who are at significant risk from fire - consider the steps necessary to reduce the risk.

■   Identify any structural features that could promote the spread of fire - consider, where possible, steps to reduce the potential for a fire to grow.

■   Monitor during maintenance and refurbishment periods the introduction of ignition sources and combustible materials into the workplace.

■   Mitigation measures in place for persons in and around premises.

■   Fire prevention policies and practices.

■   Control methods for fire spread.

■   Fire detection / warning.

■   Means of escape.

■   Means of fighting fire.

■   Level of fire safety training given to employees.

■   Emergency plan in place.

Outcomes of incidents

HUMAN HARM

Fire has the potential to kill or injure people and in 2004 there were 508 deaths due to fire and 14,600 non-fatal casualties.

Fire will in essence kill or injure people due to the effects of heat and smoke. The heat from a fire will directly affect people by physical burns, inhalation of superheated air which can scorch the lining of the lungs and heat stroke which can cause physical collapse. Smoke inhalation causes short term damage, long term damage or death. Smoke is a combination of toxic gasses and fuel particles. The toxicity of the fire gasses will often cause an individual to collapse in a fire and death due to asphyxiation will often follow. If the individual survives the fire, the toxicity of the gasses will often cause some form of damage to the lungs and respiratory tract with long term damage which is often un-repairable.

See also - toxicity on the gasses involved - later in this element.

Fires can also cause harm to people indirectly, for example with physical injuries caused whilst evacuating a fire situation or post event stress and anxiety.

We need to consider what type of harm can happen to people as part of the risk assessment:

■   Can people be trapped by fire?

■   Will they find it very difficult to escape?

■   Will the building structure or system prevent smoke travel?

■   Are people at risk of being exposed to a smoke inhalation hazard?

To understand the potential human harm in the event of a fire you need to understand how fire, heat and smoke could spread within the premises.

LEGAL AND ECONOMIC EFFECTS ON THE ORGANISATION

If a fire occurs it may indicate that insufficient action has been taken to prevent a fire and that a breach of legislation led to the fire. This means that for each fire incident that occurs there is a real prospect of prosecution of the responsible person for failing to meet their duties under the Regulatory Reform (Fire Safety) Order (RRFSO) 2005.

Article 9 of the RRFSO places a legal requirement for the 'Responsible Person' to make a suitable and sufficient assessment of the risks to which relevant persons are exposed for the purpose of identifying the 'General Fire Precautions' needed to comply with the requirements and prohibitions imposed by the order.

By conducting a suitable and sufficient fire risk assessment it should ensure compliance with the law. However, from a more practical viewpoint a good fire risk assessment should reduce the chances of fire starting in the premises. By doing so, the chances of harm to people, property or the environment will be reduced and as such the potential for financial losses will be lower.

IMPACT ON OVERALL RISK MAGNITUDE

Risk is determined to be a combination of likelihood and consequence of a fire. Though fires are not frequent events taken from a local perspective, considered nationally there is a real potential for them to occur. Hazards that may lead to a fire are an integral part of workplaces. Of the many hazards that an organisation faces few have the potential to cause such large scale harm as a fire. It therefore represents the highest potential consequence in most workplaces, with the possibility of multiple deaths, large scale damage and potential business disruption.

This means that although harm may be rare, when considered with its consequences it remains the highest magnitude of risk potential for many organisations. With this knowledge many organisations have taken a great deal of preventive and precautionary action to minimise the likelihood of a fire starting and the consequences if one should.

As indicated above, a well conducted fire risk assessment coupled with a good fire risk management system will assist with lowering the level of risk within a premises. The first outcomes from a fire risk assessment are often improved fire prevention measures and as such the likelihood for a fire to start is often reduced. This in turn will lower the business loss risks to a commercial company and if they are still unfortunate enough to suffer a fire, good fire management policies should have also considered the business continuity issues that follow on from a fire.

Differences between different types of fire incident

INJURY ACCIDENT

"an unplanned, uncontrolled event involving fire which led to or could have led to direct physical harm to people"

The possibility of death as a result of a fire is a real prospect. It is hoped however, that by having completed a comprehensive fire risk assessment and having implemented good preventive and protective measures that this possibility is removed as far as is reasonably practicable. Other injuries may be acute in nature and recognised immediately, such as burns or the effects of smoke inhalation. Acute injuries could also be caused as a result of a fire evacuation such as crush injuries, broken bones, sprains, or cuts, as a result of falls whilst evacuating.

ILL-HEALTH

"harm to a person's health caused by fire or the products of a fire"

Smoke inhalation can be damaging to people's health resulting in severe difficulties in breathing. If for example a member of staff was to inhale 'hot smoke', even for a short period of time, the heat involved can scorch the linings of the lungs with very serious consequences. Even a small amount of smoke inhalation can trigger off ill-health due to existing illnesses. For example, a person who suffers with asthma will be quickly affected by smoke and this could result in an increase in severity of their illness, with higher levels of sickness as a result.

Smoke inhalation over long periods of time can cause ill health issues such as respiratory difficulties. This however, is a greater problem to fire fighters who are confronted by smoke on a regular basis. In today's Fire and Rescue Service even this should not be an issue due to the increased use of breathing apparatus at fire incidents. Members of staff within companies should not be exposed to smoke from fires over a prolonged period of time and as such the cumulative effects on health should not be present.

DANGEROUS OCCURRENCE

"an accident not resulting in personal injury reportable to the enforcing authority"

The schedule to the Reporting of Injuries, Disease and Dangerous Occurrences Regulations (RIDDOR) 1995 lists events which must be formally reported to the relevant enforcement agency, the Health and Safety Executive or Local Authority. The main specified event relating to fire or explosion specified in Schedule 2 of RIDDOR is:

"An electrical short circuit or overload attended by fire or explosion which results in the stoppage of the plant involved for more than 24 hours or has the potential to cause the death of a person".

NEAR MISS

"an accident with no apparent loss"

A near-miss is an incident with the potential to cause harm, but where there is no measurable loss. In relation to fire this may include such events as observing someone smoking near a flammable substances store or observing overloaded electrical equipment that is beginning to give off smoke.

It is important to analyse any near miss to assess the potential of the event, had circumstances been different. This will enable corrective action to be put in place to prevent a re-occurrence of the incident.

FIRE DAMAGE ONLY

"an accident were no people were harmed but equipment or materials are harmed by a fire"

The reason why this is the only result of the event may be because of prompt extinguishing and evacuation arrangements or the amount of fuel available to the fire was limited. Whichever is the case it is important to report and consider such events for their potential to re-occur and the positive or negative lessons that can be learnt concerning prevention and control of fire.

Typical ratios of incident outcomes

Information taken from the National Fire Statistics 2004 gives the following facts. The figures shown are given based on all premises except dwellings.

Total number of fires	37,600	
Deaths	55	Equates approximately to 1.5 death per 1,000 fires Approximately 68% due to smoke inhalation
Non fatal injury (non-fire fighter)	1,500	Equates approximately to 40 injuries per 1,000 fires

Figure 6-5: Typical ratios of incident outcomes (non-fire fighter). *Source: National Fire Statistics 2004.*

Total number of fires	37,600	
Deaths (fire fighter)	3	
Non fatal injury (fire fighter)	137	Equates approximately to 4.4 injuries per 1,000 fires

Figure 6-6: Typical ratios of incident outcomes (fire fighter). *Source: National Fire Statistics 2004.*

Pyramid model of incident outcomes

Figure 6-7: Accident ratio study. *Source: Frank Bird.*

The principle applies equally to fires as it does to other situations. The organisation, site or department that has most near misses is therefore, in the long run, likely to have the most serious losses.

The other lesson to be learned is that if you tackle the causes of minor consequence fires, you will automatically reduce the serious loss fire causes as well.

Utility and limitations of accident ratios in accident prevention

It is often the practice to use statistics as a factor when assessing the probability of an incident to occur. For example when looking at the fire statistics for accidental fires it can seen that the most common single material to be first ignited is electrical insulation, followed by waste paper and cardboard. From this information we can surmise that targeted fire prevention methods aimed at electrical equipment, waste materials, paper and cardboard stocks would dramatically reduce the probability of a fire starting in the workplace.

The accident ratio triangle illustrates the importance of identifying fire events leading to a range of consequence, including near misses as these provide an opportunity to prevent fires with major loss. An organisation will not always get early warning of a risk of fire through the occurrence of near misses, the first fire experienced may be a major one.

Fire risk assessment process

IDENTIFYING FIRE HAZARDS

Sources and form of harm

The first step is to identify the hazards. This can be done in many ways, with the most common method being to walk around the premises with some form of prompt checklist to follow. Whatever method is used it will need to highlight all of the hazards associated within a premises including those created as a result of a specific task. For complex premises, it may be preferable to break the premises down into manageable parts. If this is done it is critical that a holistic overview is taken at the end of the process to verify that hazards found in one area do not impact on another area.

- Identify sources of ignition - smoker's materials, naked flames, heaters, hot processes, cooking, machinery, boilers, faulty or misused electrical equipment, lighting equipment, hot surfaces, blocked vents, friction, static electricity, metal impact and arson.
- Identify sources of fuel - flammable liquids, flammable chemicals, wood, paper and card, plastics, foam, flammable gases and liquefied petroleum gases (LPG), furniture, textiles, packaging materials, waste materials including shavings, off cuts and dust.

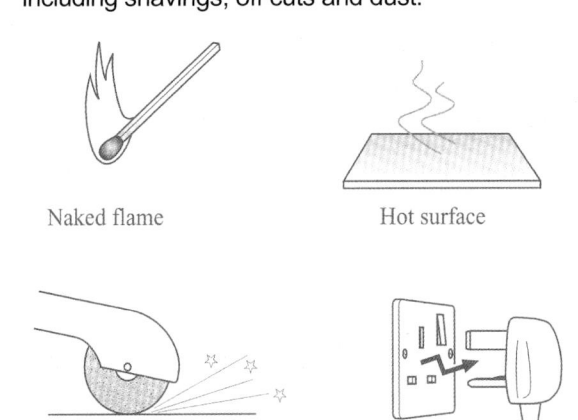

Naked flame

Hot surface

Mechanically generated sparks

Electrically generated sparks

Figure 6-8: Sources of ignition.   *Source: HSE Books.*   Figure 6-9: Pallets and flammable liquids - fuel sources.   *Source: FST.*

- Identify sources of oxygen - natural ventilation, doors, windows, forced ventilation systems, air conditioning, oxidising materials, oxygen cylinders or piped oxygen systems.
- Identify structural features that may spread fire - combustible wall, floor or ceiling linings, open voids, open ducts, breeches in fire resistance.

Figure 6-10: Sources of oxygen.   *Source: FST.*   Figure 6-11: Steel shutter to prevent fire spread.   *Source: FST.*

When hazard spotting we need to concentrate on significant hazards and may choose to ignore the trivial ones. When considering the lists above, we need to consider the interaction of each factor. For example, we need to identify ignition sources, but a significant hazard would be a heat source that has sufficient heat output to ignite the fuels that are present. This of course is dependent upon the format that the fuel is presented in at the premises.

As an example, we would not need to be concerned about a spark from an electric motor if the fuel in the area was thick timber sections, but if the timber was in the form of a fine dust the spark may be sufficient to cause an explosion.

Roles of inspections

Inspections provide an opportunity to observe the workplace conditions and actions of people that present a hazard and increase the risk of a fire. In conducting a risk assessment it is essential that it not be limited to a 'table top exercise' looking at building plans and that a significant amount of time is spent assessing the actual situation.

Inspections, as part of the management control of hazards, involve examination of the workplace or items of equipment in order to identify fire hazards and determine if they are effectively controlled. Four different types of inspection are common:

- General workplace inspections - carried out by local first-line managers and employee representatives.
- Statutory inspections (thorough examination) of equipment, e.g. boilers, electrical equipment - by specialist Competent Persons.
- Preventive maintenance inspections of specific (critical) items - carried out by maintenance staff.

■   Pre use 'checks' of equipment, e.g. gas cutting equipment - carried out by the user.

They represent an opportunity to identify a new hazard or the changed state of a control, which may have a significant bearing on a risk assessment that has previously been conducted.

Job/task analysis

Job / task analysis is the systematic identification of all the fire hazards and prevention measures appropriate to a particular job or area of work activity, and the behavioural factors which most significantly influence whether or not these measures are taken. The approach is diagnostic as well as descriptive.

Analysis can be:

Job based	Task based
Maintenance engineer	Electrical welding activities
Chemical process operator	Decanting flammable liquid stocks for use

The results of the analysis can be used to correct existing controls and to improve such things as emergency procedures, reporting of information and the layout of work areas. The process of job analysis needs to be carried out methodically through a series of steps and the whole analysis should be documented.

Legislation

Information about hazards can be obtained by considering a range of legislative documents including approved codes of practice (ACOPs) and guidance notes. For example, the ACOP L135 'Storage of Dangerous Substances' gives information about the storage of dangerous substances in compliance with the Dangerous Substances and Explosive Atmosphere Regulations (DSEAR) 2002. The 11 guides to the RRFSO give guidance on fire hazards and fire safety standards in certain premises types.

Manufacturers' information and safety data sheets

Manufacturers are required to provide equipment and product health and safety information. Information provided in relation to articles and substances must be relevant and kept up to date. Those that design, manufacture, supply or install have a duty to inform of any issues determined through research that may affect the users, including product liability aspects, safe operation and use of instructions. Suppliers provide material safety data sheets that can then be considered when carrying out a risk assessment.

Fire statistics prove useful information on past incidents that have occurred. This information can be helpful in assisting in the decision making process on hazards and risks.

Incident Data

When identifying the fire hazards within premises, it is often considered good practice to look at what has happened in the past. If information or statistics are available on fire incidents in the specific industry or on the specific task that is being assessed, then these can give a very good indication as to the possible hazards that may materialise and their probability that they will lead to a fire. If specific information is not available, either within the company or from within the industry, then it is worth considering national fire statistics. These statistics are available via the Office of the Deputy Prime Minister (ODPM) website www.odpm.gov.uk.

See also - Causes and prevention of fire for extracts from these statistics - in Element 3.

IDENTIFYING POPULATION AT RISK

As part of the fire risk assessment process the population at risk should be identified in the event of fire. Consideration should be given to their location, work activity, their knowledge of the premises and the hazards within, capability and any disabilities which might put them at risk in a fire situation

Operatives

Operatives and general staff will be found at various places and undertaking various tasks within premises. These people and their methods of work need to be considered to see if any particular people are at greater risk than others. This may be the case due to the location of the individual, the task they are undertaking, or the type of materials they have to work with.

Maintenance staff

Maintenance staff, by nature of their job, are often working alone and in remote areas of a building. They may be working with equipment or materials which create fire hazards. As such we need to ensure that when working in such areas these staff can still be warned of any fire and that their escape routes will not be compromised.

Cleaners

Cleaners often work in buildings out of normal hours and are often contract workers. The building will often be secured out of normal working hours and it is important to ensure there are sufficient means of escape still available to them at all times. There is a joint responsibility for those who work in host premises, placed on their employer and the responsible person for the premises to ensure the cleaners know the building layout, its exit routes and fire routine system. Discussions should have taken place between all 'responsible persons' to ensure that all factors have been considered.

Contractors

Arrangements for contractors need to be clearly established. Work to be done must be clearly defined and limitations such as where the work should be carried must be understood by all involved, and only controlled deviation allowed. Typical issues may include consideration of location of equipment or materials, and fire safety arrangements to include use of hot work permits as necessary.

Visitors and members of the public

Before visitors or members of the public are allowed into the workplace, issues of access, use of facilities, arrangements for escort or accompaniment must be established. Other issues include arrangements for safe evacuation. Certain workplaces such as places of sport have specific legal requirements placed on the duty holder, such as the appointment of fire wardens. Visitors and members of public are to a degree an unknown factor. Due to the diversity of people involved you have no knowledge of their appreciation of fire safety matters, or their potential reactions in the event of a fire. *Refer back to the issues discussed in the previous element on human nature and it should be clear that both public and visitors can be at risk due to their actions, or possibly their lack of action!*

EVALUATING RISK AND ADEQUACY OF CURRENT CONTROLS

Likelihood of harm and probable severity

Once the fire hazards and the type of people who may be affected by fire have been identified, the next stage is to evaluate the actual risks involved. We need to consider the probability of the fuels present being ignited by the ignition sources present, plus the severity of any fire that could develop. Consideration will need to be given to any additional oxygen sources (possibility of oxidising agents in store or use) present and if so what effect they will have on the ignitability of materials or the speed of fire spread.

The premises will need to be reviewed to determine if there are any unsatisfactory structural features such as open plan stairwells, vertical shafts that are not fire stopped and so on. This requires a judgment on each hazard to decide, realistically, what is the most likely outcome and how likely is this to occur. It may be a matter of simple subjective judgement or it may require a more analytical, technique depending on the complexity of the situation.

It is worth noting that the same hazard located in a different area of the premises, will give a different outcome with a higher or lower level of severity. As an example, a data projector used for presentations which goes on fire whilst mounted on the ceiling in a lecture room will probably only cause a localised fire with disruption to the building occupants. However, a similar data projector that goes on fire when mounted to the ceiling above the final exit door of the building means that the outcome effects on the means of escape will be far greater.

Qualitative and semi-quantitative risk ranking

Qualitative

Qualitative is defined as 'observations that do not involve measurements and numbers'. Qualitative risk assessment is the most common and in the majority of cases the most appropriate method of fire risk assessment. The general method used is to follow a series of prompt questions on a fire risk assessment form, observe the premises, processes and people and then to make personal judgement calls on the findings and resultant required actions. The individual making the assessment may refer to standards, guides and other information to assist them in making the required judgements.

Semi-quantitative

The semi-quantitative risk assessment system normally involves assigning a numeric value to hazard severity and likelihood of occurrence. These values are typically in the range of 1 to 5 where the higher number reflects greater severity or likelihood. The risk factor is the product of the multiplication of the value assigned to hazard and likelihood.

The standard two sided assessment process may be too simple for fire safety issues and a third element, effects of people involving a numerical score, may need to be added.

Residual risk

As detailed earlier in the note the steps taken whilst undertaking a fire risk assessment are:

Step 1	Identify potential fire hazards
Step 2	Decide who may be in danger, and note their locations.
Step 3	Evaluate the risks and adequacy of current control measures.
Step 4	Record significant findings and action proposed.
Step 5	Action plans to carry out any necessary improvement measures.
Step 6	Keep assessment under review.

As part of step 5 a look back at the hazards found is done and where possible measures are taken to reduce risks and remove hazards.

■ Reduce sources of ignition - Remove unnecessary sources of heat or replace with safer alternatives, ensure electrical fuses etc are of the correct rating, ensure safe and correct use of electrical equipment, enforcing a 'hot work' permit system, safe smoking policy, arson reduction measures.

- Minimise potential fuel for a fire - Remove or reduce amount of flammable materials, replace materials with safer alternatives, ensure safe handling, storage and use of materials, safe separation distances between flammable materials, use of fire resisting storage, repair or replace damaged or unsuitable furniture, control and removal of flammable waste, care of external storage due to arson, good housekeeping.

- Reduce sources of oxygen - Close all doors and windows not required for ventilation particularly out of working hours, shutting down non-essential ventilation systems, not storing oxidizing materials near heat sources or flammable materials, controlling the use of oxygen cylinders and ensuring ventilation to areas where they are used.

Figure 6-12: Poorly controlled smokers materials.   *Source: FST.*   Figure 6-13: Controlled storage of flammable liquids. *Source: FST.*

- Reducing unsatisfactory structural features - Remove, cover or treat large areas of combustible wall and ceiling linings, improve fire resistance of workplace, install fire breaks into open voids.

In addition consideration is given to existing fire safety measures in the workplace and what possible improvements may be made to reduce the risk - means of escape, means for protecting the means of escape, fire detectors, fire alarms, fire fighting equipment, signage, emergency lighting, policies, procedures and training.

Fire detection and warning

- Can fire be detected quickly enough to allow people to escape?
- Can means of warning be recognized and understood?
- Do staff know how to operate system?
- Will staff know what to do if alarm operates?
- Are fire notices around workplace?

Means of escape

- How long will it take for people to escape once they are aware of fire?
- Is this time reasonable?
- Are there enough exits?
- Are exits in right places?
- Do you have suitable means of escape for all people including the disabled?
- Could a fire happen that would affect all escape routes?
- Are escape routes easily identifiable?
- Are exit routes free from obstructions and blockages?
- Are exit routes suitably lit at all times?
- Have staff been trained in the use of the escape routes?

Means of fighting fire

- Is the fire fighting equipment suitable for risk?
- Is it suitably located?
- Is it signed where necessary?
- Have people been trained to use equipment where necessary?

Maintenance and testing

- Do you regularly check all fire doors, escape routes, lighting and signs?
- Do you regularly check all fire fighting equipment?
- Do you regularly check all fire detectors and alarms?
- Do you regularly check any other equipment provided to help means of escape arrangements?

- Are there relevant instructions to staff re maintenance and testing?
- Are those who carry out maintenance and testing competent?

Fire procedures and training

- Do you have an emergency plan?
- Does the emergency plan take account of all reasonably foreseeable circumstances?
- Are all employees familiar with the plan, trained in its use and involved in testing it?
- Is the emergency plan made available to staff?
- Are fire procedure notices clearly indicated throughout the workplace?
- Have you considered all people likely to be present?

Having identified necessary improvements the actions are transferred to an action plan and steps taken to implement them. Once this action of removing hazards and improving controls has taken place, the level of risk that remains is known as the residual risk. If the hazard is not eliminated there will always be a residual level of risk and it should be as low as is reasonably practicable. It is important to check that after the improved controls have been put in place they have had the desired effect and that the risk has been reduced to the residual level anticipated. In time this will be further reviewed to determine its effectiveness and opportunity for further improvement.

Acceptable/tolerable risk levels

Societal standards change and risk acceptability reduces each year within Europe. Legislation places a general duty to reduce the level of risk so far as is reasonably practicable. The standard "practicable" places a duty on the responsible person to use any new improvements in technology.

Use of guidance

When making a judgement as to whether controls are adequate care has to be taken to consider relevant guidance. This can be in the form of guidance to legislation, Health and Safety Executive (HSE) guidance documents, industry standard guidance and relevant British (European EN) Standards.

The judgement on what is acceptable / tolerable is a very important decision. The guidance can assist in this decision making process by comparing standards in the premises to the standards in the guidance. Where there is a variance it will be necessary to justify the use of an alternative practice or standard.

Sources and examples of legislation applying controls to specified hazards

Duties to consider specific aspects are found in the schedules to the RRFSO and DSEAR.

General control hierarchy

When considering the fire hazards/risks that are present in a premises and the need to reduce these hazards/risks to as low a level as possible, it is often helpful to refer to the general control hierarchy.

E liminate	-	the material/substance or work practice.
R educe	-	the use or frequency or substitute - for a lesser hazard or change the physical form.
I solation	-	the hazard from workplace by use of fire resistant storage.
C ontrol	-	at source, i.e. fume, dust extraction, totally enclose.
P PE	-	anti static clothing, conductive operator footwear.
D iscipline	-	follow the rules, obey signs and instructions.

Prioritisation based on risk

Once the improvement measures required as a result of the fire risk assessment have been highlighted, it may be desirable to allocate some form of measurement of scale of risk e.g. high, medium, low, or a score of 1 - 5. In this way the action plan timescales can be written based on these levels of risk.

Distinction between priorities and time scales

Significant risks are of high priority, but the need to establish realistic time scales is also important. Whilst some risks may carry a high priority due to their significance the remedies to improve the risks may take a significant time to improve. However, it may be necessary to put in intermediate actions to reduce the risk while waiting for a longer term solution to be completed. Often it is possible to carry out some changes in the short and medium term to reduce the likelihood of a loss and remove the need to give everything considered a high priority, unrealistic timescale for completion.

ACTION PLANS TO CONTROL RISKS

One of the outcomes of the fire risk assessment must be to produce action plans detailing the necessary improvement measures that are needed. These will provide a clear list of actions to improve fire risk and the timescales available to complete them. This helps to organise resource and direct effort where it will have the greatest effect. It provides an opportunity to identify those items that have or have not been completed, as well as a well structured status summary for communication to those that need to take action.

RECORDING SIGNIFICANT FINDINGS

Format

Employers with five or more employees are required to record the *significant findings* of their risk assessments in writing.

It should be noted that there are many forms and systems designed for recording assessments and while these may differ in design, the methodology broadly remains the same. Suitable headings of the record are:

Significant Hazard	People at risk	Existing control measures	Further action needed	Projected completion date	Responsible person

Information to be recorded

A written record of the findings of the risk assessment, together with details of any people identified as being at particular risk must be made. A record of action taken must also be kept.

REVIEWING THE FIRE RISK ASSESSMENT

Reasons for review

The fire risk assessment and fire safety measures (including procedures) must be reviewed on a regular basis, if not reviewed for other reasons.

Reasons that would cause a review to take place are:

- Incidents, such as if a near miss or a fire occurs.
- Incidents by a third party.
- Changes are proposed or made to a work process / activity / substance.
- Changes are proposed or made to the workplace.
- Changes to numbers or type of people (e.g. young persons, those with disability) in the workplace.

Incidents
Incidents that occur may confirm or challenge the level of risk and suitability of controls in place and as such have a major bearing on a risk assessment that has been conducted.

Changes to process
Changes to a work process / activity may have altered the level of risk.

Equipment
Changes to workplace layout or equipment may amend the escape routes and could reduce fire standards. Reassessment is necessary.

Substances used

If new substances are proposed or introduced into the workplace, then a reassessment may be necessary due to increased risks.

Staff

Changes to number or type of people (e.g. young persons, those with disability) present will require a new assessment.

Legislative changes

If an amendment to legislation or new legislation is made, the fire risk assessment will need to be reviewed to ensure compliance.

Elapse of time

The assessment needs to be reviewed at regular periods. Many organisations carry out annual re-assessment, though legislation such as Control of Substances Hazardous to Health Regulations cites a maximum period of five years. This may be seen as sufficient in a very stable organisation where little changes.

CRITERIA FOR A 'SUITABLE AND SUFFICIENT' RISK ASSESSMENT

In order to be suitable and sufficient a risk assessment must:

1) Identify all the hazards and evaluate the risks from those hazards, taking into account current legal requirements.

2) Record the significant findings.

3) Identify any group of employees or single employees who are especially at risk.

4) Identify others who may be specifically at risk e.g. members of the public, visitors.

5) Evaluate existing controls, stating whether or not they are satisfactory, and if not, the action necessary.

6) Evaluate the need for further controls including information, instruction and training.

7) Be carried out by competent person(s) who have the necessary experience or training in hazard identification and carrying out risk assessments, knowledge of the process or activity, and good communication and reporting skills. The individual(s) should have the right attitude to carry out the task, the ability to interpret legislation and guidance, and technical knowledge of the plant or equipment involved.

The content of a training course for staff who are to assist in carrying out risk assessments should include:

- The legal requirements with respect to risk assessment.
- The process of identifying hazards and evaluating risks.
- The identification and selection of appropriate control measures.
- The awareness of the individual's own limitations and the occasions when specialist assistance might be required.
- Accessing sources of information such as ACOPs and in-house information including accident records.
- Report writing skills.
- The interpretation of regulations and standards.
- The means available for disseminating the outcomes of the assessment.

6.2 - Environmental impact of fire

Fire is very damaging not only to the premises on fire, but to its surrounds. Under Article 16 of the RRFSO there is the legal duty to 'Mitigate the effects of fire'. This statement could be very wide ranging in its effects and consideration must not only be given to the effects of a fire on the premises where the fire may start but also the effects on neighbouring properties, the local community and the environment. Environmental legislation is diverse and applies to many different types of premises, ranging from those with minor to major impact potential to the environment. As an example:

- Control of Major Accident Hazards Regulations 1999 (COMAH) which came into force on 1 April 1999 as amended by the Control of Major Accident Hazards (Amendment) Regulations 2005 from 30 June 2005.

COMAH exists to ensure that businesses:

- Take all measures necessary to prevent major accidents involving dangerous substances; and
- Limit the consequence to people and the environment of any major accidents which do occur.

The Environmental Liability Directive came into force in April 2004. It is aimed at preventing environmental damage by forcing industrial polluters ("operators") to pay prevention and remediation costs.

Companies carrying out other, less harmful, activities will be held liable when damage to protected habitats and species has been caused by their fault or negligence.

It follows that if a fire or explosion, is proven to be caused by a preventable action or event, and environmental damage has occurred, the 'responsible person' may be deemed liable to prosecution for the failure to have appropriate mitigation measures in place.

Once a large fire has started it is very difficult to extinguish and some environmental damage will occur. The only way to prevent such events is to invest in fire prevention and control measures to reduce the possibility of a major event occurring.

Sources of pollution in the event of a fire

Pollutants may escape from the site into the water ecosystem by a number of pathways. These include:

- The site's surface water drainage system, either directly or via off-site surface water sewers.
- Direct run-off into nearby watercourses or onto ground, with potential risk to groundwaters.
- Via the foul drainage system, with pollutants either passing unaltered through a sewage treatment works or affecting the performance of the works, resulting in further environmental damage.
- Through atmospheric deposition, such as vapour plumes.

TOXIC AND CORROSIVE SMOKE

It is a well known fact that fires can give off large quantities of toxic fumes and smoke, which may also contain pollutant particles such as asbestos. They will be carried by prevailing winds and can cause harm over long distances before they disperse and as they fall to the ground in rain water.

Smoke

Smoke consists of small particles or partially burnt, carbonaceous materials. The colour, size and quantity of these particles will determine the thickness of the smoke; water vapour also thickens smoke.

Smoke and its by-products are normally very corrosive and as such can cause long term damage to buildings and materials unless cleaned correctly following a fire.

Figures 6-14 & 6-15: Potential pollution from fire run off.   Source: Royal Chiltern Air Support Unit.

Production of toxic and corrosive gases

For plastics that contain the elements carbon and hydrogen, or carbon, hydrogen and oxygen, the main toxic gas to be expected when plastics are involved in a fire is carbon monoxide (CO) which is formed when organic materials are burned with reduced availability of oxygen. This gas is a well-known hazard to fire-fighters as it is produced in the combustion of traditional materials. It is odourless, colourless and a very toxic gas. Certain plastics containing only carbon, hydrogen and oxygen, such as some types of phenol-formaldehyde resin, can produce products of combustion other than carbon monoxide at lower temperatures ($470^{\circ}C$). Under these conditions phenol is produced, which would present an additional toxic hazard.

For plastics that contain nitrogen in addition to carbon, hydrogen or oxygen, we may expect nitrogen compounds in the fire gases; these are also potentially very toxic. Such nitrogen-containing materials are cellulose nitrate, nylon, polyurethane foams, melamine-formaldehyde plastics, urea-formaldehyde plastics, ABS (acrylonitrile-butadiene-styrene), some epoxy resins and nitrile rubbers. In the gases given off as a result of fire, the nitrogen of these materials may well appear as nitrogen dioxide, hydrogen cyanide and various organic nitriles (cyanides), all of which are toxic, especially so in very small quantities in the case of cyanides and nitriles.

Polyurethane foams have been extensively investigated and hydrogen cyanide and carbon monoxide are present in appreciable amounts in the fire gases. Less is known concerning the other materials, but there is a strong possibility that toxic nitrogenous compounds are present in the fire gases. In the case of cellulose nitrate, hydrogen cyanide, nitrogen oxide and carbon monoxide are known to be present.

Chlorine is an element present in polyvinyl chloride (PVC) and certain related co-polymers, in neoprene and in certain types of self-extinguishing fibre-glass polyester resin. In the case of PVC almost all the chlorine of the molecule appears as hydrogen chloride gas in the fire gases. This is both toxic and corrosive; it has a very sharp penetrating odour and with water it will form solutions of hydrochloric acid, which is also corrosive. Apart from corroding many metals, the acid may cause long term changes in alkaline mortar.

Ferroconcrete may be much less affected; nevertheless, copious washing after incidents involving PVC is very desirable. Other chlorine-containing polymers may also give hydrochloric acid gas and possible other chloride-containing toxic compounds as well.

Fluorine-containing polymers are PTFE (Polytetrafluoroethylene), certain related materials, such as "Kel-F" and a series of synthetic rubbers, sometime known as "vitrons". If these materials are overheated, toxic products from decomposing fluorinated materials are produced .

Carbon is present in most combustible materials and is readily oxidized under fire conditions to form carbon dioxide (CO_2) if oxidation is complete, or carbon monoxide (CO), if the supply is limited, e.g. in an enclosed space or in the middle of a fire.

Carbon monoxide

It is an odourless and invisible gas, which can quickly spread throughout a burning building because its density is similar to that of air. When it is inhaled and absorbed into the bloodstream it chemically joins with haemoglobin molecules (red blood cells) which would normally be carrying oxygen around the body. Carbon monoxide will bond with the haemoglobin, with an affinity of 200 times that of oxygen and is known as a *chemical asphixiant*. A victim may be breathing in some oxygen, but it will not be transported readily to the body tissues by the haemoglobin, and if exposure is prolonged this will result in suffocation.

Because the gas is odourless, its presence may be undetected until collapse is imminent. The first effects are similar to alcoholic intoxication, and the victim's judgement is impaired. Continued inhalation causes unconsciousness and eventually death. 1.3 per cent in the air is enough to cause unconsciousness after two or three breaths, and as little as 0.32 per cent can cause death in thirty minutes.

Carbon dioxide

Is slightly toxic, but acts mainly as an asphyxiant by lowering the proportion of oxygen available for breathing. It is odourless and invisible and though it is more dense than air, it can spread throughout a building quickly under fire conditions.

Concentrations of up to three per cent in air cause rapid deep breathing which enhances the body's intake of other toxic gases. At 5 per cent breathing becomes laborious and at 9 per cent the victim is likely to lose consciousness within a few minutes. Concentrations of twenty per cent in air can kill in twenty minutes.

Run-off of contaminated fire-fighting water

The Fire and Rescue Service will have to use a large quantity of water when fighting a major fire. The excess water that is not turned into steam by the heat from the fire will disperse or 'run off' from the fire ground area as polluted water. The Fire and Rescue Service, in conjunction with the environmental agencies, will attempt to prevent any environmental damage as a result of such factors and fire fighting operations may even be amended or halted when the risk assessment of environmental damage is high. As discussed earlier, if the water run off does create pollution, and the fire was due to a preventable incident or event, a challenge on the mitigation measures taken by the site operator may be made.

In anticipating the likelihood of contaminated water it would not be uncommon for sites such as chemical works or fuel storage facilities to install a system of water interception so that the majority of the water does not make its way into storm water drains and cause extended pollution.

Legal obligations related to environmental protection in the event of a fire

Responsible persons for the premises affected will have legal obligations for environmental protection in the event of a fire. It is illegal to allow contaminated water to enter surface drains and a company can be fined up to £20,000 for allowing such an event and any associated restoration of the environment costs.

Role of the environment agency in the event of a fire

The Environment Agency along with the Fire and Rescue Service will work together at the scene of a fire to try to prevent pollution resulting from fires. It is normal for the Environment Agency to give advice to the Fire and Rescue Service on pollution control and generally the Fire and Rescue Service will follow this advice unless there are strong operational safety reasons for not doing so.

Water Resources Act (WRA) 1991

The Water Resources Act (WRA) 1991 principally deals with the abstraction of water from rivers, boreholes etc. It is normal practice that a licence is required to abstract water other than in certain conditions.

However, under regulation 32(2) if the water is required for anything to do with:

- Fire-fighting purposes.
- The purpose of testing apparatus used for those purposes or of training or practice in the use of such apparatus, then an application for a licence is not required.

Pre-planning the minimisation of environmental impact of fire

Most industrial and commercial sites have the potential to cause significant environmental harm and to threaten water supplies and public health. Spillages of chemicals and oil are obvious threats. However, materials which are non-hazardous to humans, such as milk and beer, may also cause serious environmental problems, as can the run-off generated in the event of a fire. The environmental damage may be long term and, in the case of groundwater, may persist for decades or even longer.

When pre-planning for 'minimising the effects of fire' one of the considerations therefore has to be the possibility of contaminated water running off the site and into the water course. It would be prudent to consider the following:

- Materials / chemicals on site that would act as pollutants.
- Immediate local area for natural and man made water courses, rivers, streams, lakes, ponds, canals etc.
- Position of surface water drains.
- Lie of land and expected flow pattern of any run off.
- Water table levels on site.
- Existing primary containment of materials by tank, bunding etc.
- Location of shut off valves so as to limit damage.
- Isolation of oil separators (interceptors) to protect sewers in the event of use of foam to fight fires. Fire fighting foam, which contains surfactants similar to those used in cleaning solutions, will cause oil and water to mix in separators and defeat the separation process resulting in oil discharge to sewers.
- Identification of any potential areas that could be utilised as a secondary containment reservoir for contaminated water.

As discussed earlier consideration will need to be given to potential air pollution via smoke products such as:

- Discharge of toxic chemicals into air e.g. chlorine.
- Type of smoke potentially produced and environmental damage, airborne asbestos, carbon particles etc.
- Prevailing wind conditions.
- Vicinity of other properties.
- Communication methods to local community.

Further consideration will need to be given to the possible release of flammable gasses / vapours into the air and the potential effects these may have on the local community.

Procedures for containment of fire fighting water run off

Article 13(3)(c) of the RRFSO requires the 'responsible person' to arrange any necessary contacts with the emergency services regarding fire fighting. Article 16(3) of the RRFSO requires immediate steps to be taken to 'mitigate the effects of fire', if dangerous substances are on site.

Considering these two articles along with the environmental considerations highlighted earlier, if it is proposed to store dangerous substances on a site in any quantities, discussions with the fire and rescue service prior to any change of use or potential incident is critical. It is only by having a pre-planned action plan in place and tested, that the most effective procedures for containment of water run off during a fire can be achieved.

As an example of some of the effects of this need to control run off from fires, we can consider the incident that happened at Buncefield Oil Terminal on 11th December 2005.

The statements below were issued by the Environment Agency on 21st February 2006.

- 'All the contained contaminated firewater has been safely removed and is now stored at a number of sites around the country. We will ensure that operators, on behalf of the oil companies, dispose of the contaminated water with great care.'
- 'A black oil-like liquid was recently discovered in a road drain to the north of the Buncefield site. The liquid continues to be pumped out of the drain and we have taken samples to identify the liquid. We are investigating possible drainage links to see if the groundwater has been polluted. As yet there are no signs of this. We will continue to develop our comprehensive monitoring plan as we gain more evidence and understanding of any likely pollution.'
- 'Since the incident Three Valleys Water Company have suspended use of the nearest borehole for drinking water as a precaution.'
- 'We have continued to monitor the nearby River Ver, which showed low levels of pollution after the incident, but are now registering as normal.'

Buncefield was a very large incident to say the least, but these statements give an indication of the need for preplanning to prevent environmental damage. One other factor to consider is of course that Buncefield was a registered COMAH site, and as such would have been required under licence to establish pollution control arrangements for all foreseeable occurrences.

For further information on the environmental issues for fires and water run off, it may be worth looking at the website page shown below 'Managing Fire Water and Major Spillages: PPG18'. The information on this website which is produced by the Environment Agency has been utilised in compiling this note.
www.netregs.gov.uk/commondata/acrobat/ppg18.pdf

SACRIFICIAL AREAS

The effects on the environment may be minimised by pumping run off water to a remote, designated sacrificial area where drains can be stopped and the water held. The area may also be used for other purposes on a day to day basis prior to the use of it as a sacrificial area, for example, as car parking or as a sports ground.

BUNDING OF VEHICLE PARKING AND OTHER HARD STANDINGS

Impermeable yards, roads and parking areas can be converted to temporary lagoons using sandbags, suitably excavated soil or sand from emergency stockpiles to form perimeter bunds. In the event of an incident, all drain inlets, such as gullies, within the area, must be sealed to prevent the escape of the pollutant.

PITS AND TRENCHES

Pits or trenches may be used where other methods have failed or no other method is available. Their use should be considered carefully due to the risk of groundwater contamination. If possible, a liner should be employed.

PORTABLE TANKS, OVERDRUMS AND TANKERS

Portable storage tanks made from synthetic rubber, polymers and other materials are available in a wide variety of sizes. The portability of the tanks allows them to be moved rapidly to the fire or spillage location, or to where any run-off has been contained.

Relevant Statutory Provisions

Content

There have been a number of changes to legislation, some of which do not come into force until October 2006; the significant ones relating to the **NEBOSH Certificate in Fire Safety and Risk Management** are dealt with in this element. NEBOSH do not examine on legislation until it has been in force for 6 months. Students may show knowledge of new legislation in their answers until that point; students referring to the former legislation will not lose marks until the 6 month period has passed.

Building (Scotland) Act 2003

Law considered in context / more depth in Unit 4.

Arrangement of Regulations

PART 1

Building regulations

1) Building regulations.
2) Continuing requirements.
3) Relaxation of building regulations.
4) Guidance documents for purposes of building regulations.
5) Compliance with guidance documents.
6) Building standards assessments.

PART 2

Approval of construction work etc.

7) Verifiers and certifiers.
8) Building warrants.
9) Building warrants: grant and amendment.
10) Building warrants: extension, alteration and conversion.
11) Building warrants: certification of design.
12) Building warrants: reference to Ministers.
13) Building warrants: further provisions.
14) Building warrants: limited life buildings.
15) Building warrants: late applications.
16) Applications and grants: offences.
17) Completion certificates.
18) Completion certificates: acceptance and rejection.
19) Completion certificates: certification of construction.
20) Completion certificates: offences.
21) Occupation or use without completion certificates.
22) Imposition of continuing requirements by verifiers.
23) Discharge and variation of continuing requirements imposed by verifiers.
24) Building standards registers.

PART 3

Compliance and enforcement

25) Building regulations compliance.
26) Continuing requirement enforcement notices.
27) Building warrant enforcement notices.

PART 4

Defective and dangerous buildings

28) Defective buildings.
29) Dangerous buildings.
30) Dangerous building notices.

PART 5

General

PART 6

Supplementary

Outline of key points

The Building (Scotland) Act 2003 is similar to the Building Regulations in England and Wales in as much as it sets standards for building design so as to ensure safety for the building occupants. Buildings will need to be issued with a building warrant to certify it has achieved standards.

As with the Building Regulations in England and Wales the building standard achieved does not necessarily ensure compliance with the Regulatory Reform (Fire Safety) Order (RRFSO) 2005 and a suitable and sufficient fire risk assessment must be undertaken.

As the regulations are new they incorporate modern thinking on means of escape for the disabled. The regulations discuss the issue of 'temporary refuges' as follows:

"The intention is to allow wheelchair users to wait temporarily until it is safe to use the escape stair. The spaces are not intended to be used by people to await rescue from the fire service. The speed of evacuation of people with mobility problems can be much slower than able-bodied people and it is for this reason that temporary refuge is important on escape stairs. The added benefit to the inclusion of temporary waiting spaces allows any person with impaired mobility to use the space."

Building Regulations 2000 Access to and use of buildings - Approved Document M 2004 edition

Law considered in context / more depth in Unit 4.

Arrangement of Regulations

1) Access to Buildings Other Than Dwellings.

2) Access into Buildings Other Than Dwellings.

3) Horizontal and Vertical Circulation in Buildings Other Than Dwellings.

4) Facilities in Buildings Other Than Dwellings.

5) Sanitary Accommodation in Buildings Other Than Dwellings.

6) Means of Access to and into the Dwelling.

7) Circulation within the Entrance Storey of the Dwelling.

8) Accessible Switches and Socket Outlets in the Dwelling.

9) Passenger Lifts and Common Stairs in Blocks of Flats.

10) WC Provision in the Entrance Storey of the Dwelling.

Outline of key points

Part M of the Building Regulations links in with the Disabled Discrimination Act (DDA) 1995 as they both deal with access for the disabled.

The following statement is taken from the introduction to the Part M document.

"Means of escape in case of fire: the scope of Part M is limited to matters of access to, into, and use of, a building. It does not extend to means of escape for disabled people in the event of fire, for which reference should be made to Approved Document B, Fire Safety."

Unfortunately a lot of people comply with part M, but do not review their fire risk assessment and the means of escape in case of fire. As a result there will be people in premises with disabilities, who cannot escape in the event of a fire. The Regulatory Reform (Fire Safety) Order (RRFSO) 2005 places a responsibility on the 'Responsible Person' to ensure that all persons can escape in the event of a fire.

Building Regulations 2000 Fire Safety - Approved Document B (consolidated with 2000 and 2002 amendments)

Law considered in context / more depth in Unit 4.

Outline of key points

Approved Document B is a Fire Safety document that provides practical guidance on meeting the requirements of Schedule 1 to and Regulation 7 of the Building Regulations 2000 for England and Wales. The areas covered are as follows:

- B1 Means of warning and escape.
- B2 Internal fire spread (linings).
- B3 Internal fire spread (structure).
- B4 External fire spread.
- B5 Access and facilities for the Fire Service.

The current Approved Document B provides a consolidation of the guidance previously issued in the original 2000 edition (and the subsequent amendments of it which were issued in 2000 and 2002) and is one of several Approved Documents issued by the Secretary of State that are intended to provide guidance for some of the more common building situations. However, there may well be alternative ways of achieving compliance with the requirements. Thus there is no obligation to adopt any particular solution contained in an Approved Document if you prefer to meet the relevant requirement in some other way.

B1 MEANS OF WARNING AND ESCAPE

Guidance.

Section 1:   Fire alarm and fire detection systems.

Section 2:   Dwelling houses.

Section 3:   Flats and maisonettes.

Section 4:   Design for horizontal escape - buildings other than dwellings.

Section 5:   Design for vertical escape - buildings other than dwellings.

Section 6:   General provisions common to buildings other than dwelling houses.

B2 INTERNAL FIRE SPREAD (LININGS)

Guidance.

Section 7:   Wall and ceiling linings.

B3 INTERNAL FIRE SPREAD (STRUCTURE)

Guidance.

Section 8:   Loadbearing elements of structure.

Section 9:   Compartmentation.

Section 10:   Concealed spaces (cavities).

Section 11:   Protection of openings and fire stopping.

Section 12:   Special provisions for car parks and shopping complexes.

B4 EXTERNAL FIRE SPREAD

Guidance.

Section 13:   Construction of external walls.

Section 14:   Space separation.

Section 15:   Roof coverings.

B5 ACCESS AND FACILITIES FOR THE FIRE SERVICE

Guidance.

Section 16:   Fire Mains.

Section 17:   Vehicle access.

Section 18:   Access to buildings for firefighting personnel.

Section 19:   Venting of heat and smoke from basements.

APPENDICES

Appendix A: Performance of materials and structures.

Appendix B: Fire doors.

Appendix C: Methods of measurement.

Appendix D: Purpose groups.

Appendix E: Definitions.

Appendix F: Fire behaviour of insulating core panels used for internal structures.

Appendix G: Standards referred to/other publications referred to.

Main changes in the 2000 edition

The main changes are:

GENERAL INTRODUCTION

a. Hospitals: HTM 81 can be used instead of the Approved Document.

B1

b. *Fire alarms:*

    i. the Requirement has been expanded to include fire alarm and fire detection systems;

    ii. the guidance forms a new Section 1 and has been extended to loft conversions and buildings other than dwellings.

c. *Alternative approaches:* this guidance has been moved from Section 3 to the "Introduction" and expanded.

d. *Door width:* the definition has been modified to align with that given in Approved Document M and corresponding reductions made to Table 5 "Widths of escape routes and exits".

e. *Means of escape:*

    i. Dwellings - storeys not more than 4.5m above ground level need to be provided with emergency egress windows;

    ii. Single escape routes and exits - the limit of 50 persons has been increased to 60;

    iii. Alternative escape routes - the 45° rule has been changed;

iv. Minimum number of escape routes - Table 4 has been simplified;

v. Mixed use buildings - the guidance has been modified;

vi. Door fastenings - more guidance is given;

vii. Escape lighting - changes have been made in Table 9 regarding toilet accommodation;

viii. Storeys divided into different uses - guidance has been added to deal with storeys which are also used for the consumption of food and/or drink by customers;

ix. Shop store rooms - guidance is given on when these need to be enclosed in fire-resisting construction.

B2

f. **Special applications:** guidance is given on the use of air supported structures, structures covered with flexible membranes and PTFE based materials.

B3

g. **Places of special fire hazard:** these need to be enclosed in fire-resisting construction.

h. **Compartments:** maximum compartment dimensions have been extended to single storey Schools and to the Shop/Commercial purpose group.

B4

i. **Rooflights:** separate Tables are given for Class 3 and TP(a)/(b) plastics rooflights and the provisions relating to Class 3 rooflights on industrial buildings has been modified.

B5

j. **Vehicle access:**

i. specific guidance is now included for single family houses and for blocks of flats and maisonettes;

ii. the 9m height in Table 20 has been increased to 11m.

k. **Personnel access:**

i. modifications have been made to the heights at which firefighting shafts are needed, with corresponding reductions to the 20m height in B1 (access lobbies & corridors), B3 (Table 12), B4 Diagram 40) and Appendix A (Tables A2 & A3);

ii. guidance is given regarding firefighting shafts in blocks of flats and maisonettes.

Appendix A

l. **Uninsulated glazed elements:** table A4 has been modified and extended.

m. **Notional designations of roof coverings:** bitumen felt pitched roof coverings have been deleted from table A5.

Appendix B

n. **Compartment walls:** limits are now specified on the use of uninsulated doors.

Appendix E

o. **Fire separating element:** this new definition has been added to support Sections 9 to 11 in B3.

Appendix F

p. This new Appendix gives guidance on **insulating core panels**.

Chemicals (Hazard Information and Packaging for Supply) Regulations (CHIP 3) 2002

Law considered in context / more depth in Unit 3.

Arrangement of Regulations

1) Citation and commencement.
2) Interpretation.
3) Application of these Regulations.
4) Meaning of the approved supply list.
5) Classification of substances and preparations dangerous for supply.
6) Safety data sheets for substances and preparations dangerous for supply.
7) Advertisements for substances dangerous for supply.
8) Packaging of substances and preparations dangerous for supply.
9) Labelling of substances and preparations dangerous for supply.

10) Particular labelling requirements for certain preparations.
11) Methods of marking or labelling packages.
12) Child resistant fastenings and tactile warning devices.
13) Retention of classification data for substances and preparations dangerous for supply.
14) Notification of the constituents of certain preparations dangerous for supply to the poisons advisory centre.
15) Exemption certificates.
16) Enforcement, civil liability and defence.
17) Transitional provisions.
18) Extension outside Great Britain.
19) Revocations and modifications.

Schedule 1 Classification of substances and preparations dangerous for supply.
Schedule 2 Indications of danger and symbols for substances and preparations dangerous for supply.
Schedule 3 Classification provisions for preparations dangerous for supply.
Schedule 4 Classification provisions for preparations intended to be used as pesticides.
Schedule 5 Headings under which particulars are to be provided in safety data sheets.
Schedule 6 Particulars to be shown on labels for substances and preparations dangerous for supply and certain other preparations.
Schedule 7 British and International Standards relating to child resistant fastenings and tactile warning devices.
Schedule 8 Modifications to certain enactments relating to the flashpoint of flammable liquids.

The Chemicals (Hazard Information and Packaging for Supply) Regulations (CHIP 3) 2002 apply to those who supply dangerous chemicals. They are based on European Directives, which apply to all EU and European Economic Area (EEA) Countries. The Directives are constantly reviewed and changed when necessary. When changes do occur to the Directives, CHIP is changed as well (about once a year). CHIP may be changed by amending Regulations or if there are major changes, the principal Regulations are revised.

The Regulations are designed to protect people's health and the environment by:
- Identification of the hazardous properties of materials (classification).
- Provision of health and safety information to users (safety data sheet and label).
- Packaging of materials safely.

CHIP introduces a new scheme to classify products based upon a calculation method.

Outline of key points

REGULATION 6 (1)

*'The supplier of a substance or preparation dangerous for supply **shall** provide the recipient of that substance or preparation with a safety data sheet containing information under the headings specified in Schedule 5 to enable the recipient of that substance or preparation to take the necessary measures relating to the protection of health and safety at work and relating to the protection of the environment and the safety data sheet shall clearly show its date of first publication or latest revision as the case may be.'*

The test of adequacy of the information provided in a safety data sheet is whether the information enables the recipient to take the necessary measures relating to the protection of health and safety at work and relating to the protection of the environment.

This does not mean that the safety data sheet will take the place of a risk assessment which would require specific detail of the circumstances in which the chemical is to be used.

GUIDANCE ON THE CONTENTS OF SAFETY DATA SHEETS

The headings shown here are those specified in Schedule 5 of C(HIP) 2. However, information given here is indicative of the issues to be addressed by the person compiling the safety data sheet and does not impose an absolute requirement for action or controls.

Identification of the substance/preparation and the company
- Name of the substance.
- Name, address and telephone number (including emergency number) of supplier.

Composition/information on ingredients
- Sufficient information to allow the recipient to identify readily the associated risks.

Hazards identification
- Important hazards to man and the environment.
- Adverse health effects and symptoms.

First-aid measures
- Whether immediate attention is required.
- Symptoms and effects including delayed effects.
- Specific information according to routes of entry.
- Whether professional advice is advisable.

Fire fighting measures
- Suitable extinguishing media.
- Extinguishing media that must not be used.
- Hazards that may arise from combustion e.g., gases, fumes etc.
- Special protective equipment for fire fighters.

Accidental release measures
- Personal precautions such as removal of ignition sources, provision of ventilation, avoid eye/skin contact etc.
- Environmental precautions such as keep away from drains, need to alert neighbours etc.
- Methods for cleaning up e.g. absorbent materials. Also, "Never use…."

Handling and storage
- Advice on technical measures such as local and general ventilation.
- Measures to prevent aerosol, dust, fire etc.
- Design requirements for specialised storage rooms.
- Incompatible materials.
- Special requirements for packaging/containers.

Exposure controls/personal protection
- Engineering measures taken in preference to personal protective equipment (PPE) 1992.
- Where PPE is required, type of equipment necessary e.g. type of gloves, goggles, barrier cream etc.

Physical and chemical properties
- Appearance, e.g. solid, liquid, powder, etc.
- Odour (if perceptible).
- Boiling point, flash point, explosive properties, solubility etc.

Stability and reactivity
- Conditions to avoid such as temperature, pressure, light, etc.
- Materials to avoid such as water, acids, alkalis, etc.
- Hazardous by-products given off on decomposition.

Toxicological information
- Toxicological effects if the substance comes into contact with a person.
- Carcinogenic, mutagenic, toxic for reproduction etc.
- Acute and chronic effects.

Ecological information
- Effects, behaviour and environmental fate that can reasonably be foreseen.
- Short and long term effects on the environment.

Disposal considerations
- Appropriate methods of disposal e.g. land-fill, incineration etc.

Transport information
- Special precautions in connection with transport or carriage.
- Additional information as detailed in the Carriage of Dangerous Goods by Road Regulations (CPL) 1994 may also be given.

Regulatory information
- Health and safety information on the label as required by C(HIP) 2.
- Reference might also be made to Health and Safety at Work etc Act (HASAWA) 1974 and Control of Substances Hazardous to Health Regulations (COSHH) 2002.

Other information
- Training advice.
- Recommended uses and restrictions.
- Sources of key data used to compile the data sheet.

RISK PHRASES AND SAFETY PHRASES

More useful information to help ensure the safe use of dangerous substances comes in the form of risk phrases and safety phrases. These are often displayed either on the container label or in the safety data sheet.

There are currently 48 risk phrases and 53 safety phrases. Some examples are given below and detailed information can be found in the ACOP to C(HIP) 2.

Risk Phrase		*Safety Phrase*	
R3	Risk of explosion by shock, friction, fire or other sources of ignition.	S2	Keep out of reach of children.
R20	Harmful by inhalation.	S20	When using do not eat or drink.
R30	Can become highly flammable in use.	S25	Avoid contact with eyes.
R45	May cause cancer.	S36	Wear suitable protective clothing.
R47	May cause birth defects.	S41	In case of fire and/or explosion do not breathe fumes.

Absence of hazard symbols or risk and safety advice does not mean the item is harmless.

Confined Spaces Regulations (CSR) 1997

Arrangement of Regulations

1) Citation, commencement and interpretation.
2) Disapplication of Regulations.
3) Duties.
4) Work in confined spaces.
5) Emergency arrangements.
6) Exemption certificates.
7) Defence in proceedings.
8) Extension outside Great Britain.
9) Repeal and revocations.

Outline of key points

The Confined Spaces Regulations (CSR) 1997 repeal and replace earlier provisions contained in s.30 of the Factories Act 1961.

A failure to appreciate the dangers associated with confined spaces has led not only to the deaths of many workers, but also to the demise of some of those who have attempted to rescue them.

A confined space is not only a space which is small and difficult to enter, exit or work in; it can also be a large space, but with limited/restricted access. It can also be a space which is badly ventilated e.g. a tank or a large tunnel.

The Confined Spaces Regulations (CSR) 1997, define a confined space as any place, including any chamber, tank, vat, silo, pit, pipe, sewer, flue, well, or other similar space, in which, by virtue of its enclosed nature, there is a foreseeable risk of a 'specified occurrence'.

Control of Major Accident Hazards Regulations (COMAH) 1999

Law considered in context / more depth in Unit 3.

Arrangement of Regulations

PART 1 - INTRODUCTION

1) Citation and commencement.
2) Interpretation.
3) Application.

PART 2 - GENERAL

4) General duty.
5) Major accident prevention policy.
6) Notifications.

PART 3 - SAFETY REPORTS

7) Safety report.
8) Review and revision of safety report.

PART 4 - EMERGENCY PLANS

9)   On-site emergency plan.

10)   Off-site emergency plan.

11)   Review and testing of emergency plans.

12)   Implementing emergency plans.

13)   Charge for preparation, review and testing of off-site emergency plan.

PART 5 - PROVISION OF INFORMATION BY OPERATOR

14)   Provision of information to the public.

15)   Provision of information to the competent authority.

16)   Provision of information to other establishments.

PART 6 - FUNCTIONS OF COMPETENT AUTHORITY

17)   Functions of competent authority in relation to the safety report.

18)   Prohibition of use.

19)   Inspections and investigations.

20)   Enforcement.

21)   Provision of information by competent authority.

22)   Fee payable by operator.

PART 7 - AMENDMENTS, REVOCATIONS, SAVINGS AND TRANSITIONAL PROVISIONS

23)   Amendments.

24)   Revocations and savings.

25)   Transitional provisions.

SCHEDULES

Schedule 1   Dangerous substances to which the regulations apply.

Schedule 2   Principles to be taken into account when preparing major accident prevention policy document.

Schedule 3   Information to be included in a notification.

Schedule 4   Purpose and Contents of Safety Reports.

Schedule 5   Emergency Plans.

Schedule 6   Information to be supplied to the public.

Schedule 7   Criteria for notification of a major accident to the European Commission and information to be notified.

Schedule 8   Provision of information by competent authority.

Outline of key points

The Control of Major Accident Hazards Regulations (COMAH) 1999 came into force on 1 April 1999. These Regulations implemented the Seveso II Directive (except for the land use planning requirements), and replaced the Control of Industrial Major Accident Hazards Regulations (CIMAH) 1984.

An Amendment Directive broadening Seveso II came into force in July 2005.

COMAH applies mainly to the chemical industry, but also to some storage activities, explosives and nuclear sites, and other industries where threshold quantities of dangerous substances identified in the Regulations are kept or used.

The main aim of COMAH is to prevent and mitigate the effects of a major accident involving dangerous substances, such as chlorine, LPG, explosives and arsenic pentoxide which can cause serious damage/harm to people and/or the environment. COMAH treats risks to the environment as seriously as those to people.

The main duty is to prepare a safety report which will include:

- A policy on how to prevent and mitigate major accidents.
- A management system for implementing that policy.
- An effective method for identifying any Major Accidents that may occur.
- Measures (safe plant and procedures) to prevent and mitigate major accidents.
- Information on safety precautions built into the plant when designed and constructed.
- Details of measures(fire fighting, relief systems, filters, etc) to limit the consequences of any accident.
- Information about the emergency plan for the site - this is used by the Local Authority for their off site plan.

Dangerous Substances and Explosive Atmospheres Regulations (DSEAR) 2002

Law considered in context / more depth in Unit 3.

Arrangement of Regulations

1)   Citation and commencement.
2)   Interpretation.
3)   Application.
4)   Duties under these Regulations.
5)   Risk assessment.
6)   Elimination or reduction of risks from dangerous substances.
7)   Places where explosive atmospheres may occur.
8)   Arrangements to deal with accidents, incidents and emergencies.
9)   Information, instruction and training.
10) Identification of hazardous contents of containers and pipes.
11) Duty of co-ordination.
12) Extension outside Great Britain.
13) Exemption certificates.
14) Exemptions for Ministry of Defence etc.
15) Amendments.
16) Repeals and revocations.
17) Transitional provisions.

Schedule 1.   General safety measures.
Schedule 2.   Classification of places where explosive atmospheres may occur.
Schedule 3.   Criteria for the selection of equipment and protective systems.
Schedule 4.   Warning sign for places where explosive atmospheres may occur.
Schedule 5.   Legislation concerned with the marking of containers and pipes.
Schedule 6.   Amendments.
Schedule 7.   Repeal and revocation.

Outline of key points

These new regulations aim to protect against risks from fire, explosion and similar events arising from dangerous substances that are present in the workplace.

DANGEROUS SUBSTANCES

These are any substances or preparations that due to their properties or the way in which they are being used could cause harm to people from fires and explosions. They may include petrol, liquefied petroleum gases, paints, varnishes, solvents and dusts.

APPLICATION

DSEAR applies in most workplaces where a dangerous substance is present. There are a few exceptions where only certain parts of the regulations apply, for example:

■   Ships.
■   Medical treatment areas.
■   Explosives/chemically unstable substances.
■   Mines.
■   Quarries.
■   Boreholes.
■   Offshore installations.
■   Means of transport.

MAIN REQUIREMENTS

You must:

■   Conduct a risk assessment of work activities involving dangerous substances.
■   Provide measures to eliminate or reduce risks.
■   Provide equipment and procedures to deal with accidents and emergencies.

- Provide information and training for employees.
- Classify places into zones and mark zones where necessary (to be phased in) -

Workplace in use by July 2003	-	Meet requirements by July 2006
Workplace modified before July 2006	-	Meet requirements at time of modifications
New workplace after 30 June 2003	-	Meet requirements from start.

The risk assessment should include:

- The hazardous properties of substance.
- The way they are used or stored.
- Possibility of hazardous explosive atmosphere occurring.
- Potential ignition sources.
- Details of zoned areas (July 2003).
- Co-ordination between employers (July 2003).

SAFETY MEASURES

Where possible eliminate safety risks from dangerous substances or, if not reasonably practicable to do this, control risks and reduce the harmful effects of any fire, explosion or similar event.

Substitution - Replace with totally safe or safer substance (best solution).

Control measures - If risk cannot be eliminated apply the following control measures in the following order:

- Reduce quantity.
- Avoid or minimise releases.
- Control releases at source.
- Prevent formation of explosive atmosphere.
- Collect, contain and remove any release to a safe place e.g. ventilation.
- Avoid ignition sources.
- Avoid adverse conditions e.g. exceeding temperature limits.
- Keep incompatible substances apart.

Mitigation measures - Apply measures to mitigate the effects of any situation.

- Prevent fire and explosions from spreading to other plant, equipment or other parts of the workplace.
- Reduce number of employees exposed.
- Provide process plant that can contain or suppress an explosion, or vent it to a safe place.

ZONED AREAS

In workplaces where explosive atmospheres may occur, areas should be classified into zones based on the likelihood of an explosive atmosphere occurring. Any equipment in these areas should ideally meet the requirements of the Equipment and Protective Systems Intended for Use in Potentially Explosive Atmospheres Regulations (ATEX) 1996. However equipment in use before July 2003 can continue to be used providing that the risk assessment says that it is safe to do so. Areas may need to be marked with an 'Ex' warning sign at their entry points. Employees may need to be provided with appropriate clothing e.g. anti static overalls. Before use for the first time, a person competent in the field of explosion protection must confirm hazardous areas as being safe.

ACCIDENTS, INCIDENTS AND EMERGENCIES

DSEAR builds on existing requirements for emergency procedures, which are contained in other regulations. These may need to be supplemented if you assess that a fire, explosion or significant spillage could occur, due to the quantities of dangerous substances present in the workplace. You may need to arrange for:

- Suitable warning systems.
- Escape facilities.
- Emergency procedures.
- Equipment and clothing for essential personnel who may need to deal with the situation.
- Practice drills.
- Make information, instruction and training available to employees and if necessary liaise with the emergency services.

Disability Discrimination Act (DDA) 1995

Law considered in context / more depth in Unit 5.

Arrangement of Regulations

Public authorities

1)   Councillors and members of the Greater London Authority.

2) Discrimination by public authorities.
3) Duties of public authorities.
4) Police.

Transport

5) Application of sections 19 to 21 of the 1995 Act to transport vehicles.
6) Rail vehicles: application of accessibility regulations.
7) Rail vehicles: accessibility compliance certificates.
8) Rail vehicles: enforcement and penalties.
9) Recognition of disabled persons' badges issued outside Great Britain.

Other matters

10) Discriminatory advertisements.
11) Group insurance.
12) Private clubs etc.
13) Discrimination in relation to letting of premises.
14) Power to modify or end small dwellings exemptions.
15) General qualifications bodies.
16) Improvements to let dwelling houses.
17) Generalisation of section 56 of the 1995 Act in relation to Part 3 claims.
18) Meaning of "disability".

Supplementary

19) Minor and consequential amendments and repeals and revocation.
20) Short title, interpretation, commencement and extent.

Schedule 1   Minor and consequential amendments.
  Part 1   Amendments of the 1995 Act.
  Part 2   Amendments related to disabled persons' badges.
  Part 3   Other amendments.

Schedule 2   Repeals and revocation.

Outline of key points

Discrimination is defined as:

'You discriminate against a disabled person if, on the ground of the disabled person's disability, you treat the disabled person less favourably than you treat or would treat a person not having that particular disability whose relevant circumstances, including his abilities, are the same as, or not materially different from, those of the disabled person'.

DDA has the effects of causing 'Responsible Persons' to make reasonable adjustments to premises so as to provide for access for disabled. When considering the effects of this law you need to recognise the potential that people with some form of disability may be present in the building. This factor must be considered when undertaking the fire risk assessment.

The outcome of your fire risk assessment should be that you need to ensure that all persons can escape in the event of a fire. Additional measures may need to be implemented to ensure this fact. You should not leave a disabled person in a fire refuge awaiting rescue by the fire and rescue service as this would be firstly non-compliance with the Regulatory Reform (Fire Safety) Order (RRFSO) 2005 and secondly would be deemed as discrimination against the individual under DDA.

Health and Safety (Consultation with Employees) Regulations (HSCER) 1996

Arrangement of Regulations

1) Citation, extent and commencement.
2) Interpretation.
3) Duty of employer to consult.
4) Persons to be consulted.
5) Duty of employer to provide information.
6) Functions of representatives of employee safety.
7) Training, time off and facilities for representatives of employee safety and time off for candidates.
8) Amendment of the Employment Rights Act 1996.
9) Exclusion of civil liability.

10)   Application of health and safety legislation.

11)   Application to the Crown and armed forces.

12)   Disapplication to sea-going ships.

13)   Amendment of the 1977 Regulations.

Outline of key points

1)   The HSCER 1996 came into force on 1 October 1996 and are made under the European Communities Act 1972.

2)   "Employees" do not include persons employed in domestic service in private households. Workplaces are defined as "any place where the employee is likely to work, or which he is likely to frequent in the course of his employment or incidentally to it."

3)   Where there are employees not represented by the Safety Representatives and Safety Committee Regulations (SRSC) 1977, the employer shall consult those employees in good time on matters relating to their health & safety at work. In particular they must be consulted on:

■   The introduction of any new measures which may affect their safety and health.
■   Arrangements made by the employer for appointing or nominating competent persons in accordance with regulations. 6(1) and 7(1) of the Management of Health and Safety at Work Regulations (MHSWR) 1999.
■   Any safety information the employer is legally obliged to provide to workers.
■   The planning and organisation of any health and safety training required under particular health and safety laws.
■   The health and safety consequences for employees of the introduction of new technologies into the workplace.

4)   Employers can consult either directly with employees or, in respect of any group of employees, one or more elected representatives of that group. These are referred to as "representatives of employee safety" (RES). If the latter option is chosen, then employers must tell the employees the name of the representative and the group he/she represents. An employer who has been consulting a representative may choose to consult the whole workforce. However, the employer must inform the employees and the representatives of that fact.

5)   If the employer consults employees directly then it must make available such information, within the employers' knowledge, as is necessary to enable them to participate fully and effectively in the consultation. If a representative is consulted, then the employer must make available all necessary information to enable them to carry out their functions, and of any record made under the Reporting of Injuries, Diseases and Dangerous Occurrences Regulations (RIDDOR) 1995 which relates to the represented group of employees.

6)   Representatives of employee safety have the following functions:

■   To make representations to the employer on potential hazards and dangerous occurrences at the workplace which affect, or could affect the represented employees.
■   Make representations to the employer on general matters of health and safety.
■   To represent the employees in workplace consultations with HSE or local authority inspectors.

7)   Representatives of employee safety must be given reasonable training in order to carry out their duties. Employers must meet the costs of the training and any travel and subsistence. They must also permit the representatives to take time off with pay during working hours in order for them to carry out their functions. Time off shall also be given, with pay, where this is required for any person standing as a candidate for election as a representative. Employers must also provide suitable facilities for the representatives to carry out their duties.

8)   The Employment Rights Act 1996, which gives protection against unfair dismissal or discrimination on grounds of health and safety, is amended to protect representatives of employee safety and candidates for their election.

9)   A breach of the HSCER 1996 does not confer any right of action in any civil proceedings.

10)   Ensures that certain provisions of health and safety legislation (including enforcement provisions) operate in respect of the HSCER 1996. The regulations are made under the European Communities Act 1972. Enforcement is by the enforcing authorities appointed under the Health & Safety at Work Act (HASAWA) 1974.

11)   The HSCER 1996 will apply in respect of the armed forces. However, the representatives of employee safety will be appointed by the employer, rather than elected. Furthermore, representatives in the armed forces will not be entitled to time off with pay under regulation 7.

12)   The HSCER 1996 do not apply to the master or crew of a seagoing ship.

13)   The SRSC 1977 are amended so that they now include employees of coal mines.

Health and Safety (First-Aid) Regulations (FAR) 1981

Arrangement of Regulations

1) Citation and commencement.

2) Interpretation.

3) Duty of employer to make provision for first-aid.

4) Duty of employer to inform his employees of the arrangements.

5) Duty of self-employed person to provide first-aid equipment.

6) Power to grant exemptions.

7) Cases where these Regulations do not apply.

8) Application to mines.

9) Application offshore.

10) Repeals, revocations and modification.

Schedule 1   Repeals.

Schedule 1   Revocations.

Outline of key points

2) Regulation 2 defines first aid as: '…treatment for the purpose of preserving life and minimising the consequences of injury or illness until medical (doctor or nurse) help can be obtained. Also, it provides treatment of minor injuries which would otherwise receive no treatment, or which do not need the help of a medical practitioner or nurse.'

3) Requires that every employer must provide equipment and facilities which are adequate and appropriate in the circumstances for administering first-aid to his employees.

4) An employer must inform his employees about the first-aid arrangements, including the location of equipment, facilities and identification of trained personnel.

5) Self-employed people must ensure that adequate and suitable provision is made for administering first-aid while at work.

Health and Safety Information for Employees Regulations (IER) 1989

Arrangement of Regulations

1) Citation and commencement.

2) Interpretation and application.

3) Meaning of and revisions to the approved poster and leaflet.

4) Provision of poster or leaflet.

5) Provision of further information.

6) Exemption certificates.

7) Defence.

8) Repeals, revocations and modifications.

The Schedule   Repeals, revocations and modifications.

Part I - Repeals.

Part II - Revocations.

Part III - Modifications.

Outline of key points

The Health And Safety (Information for Employees) Regulations (IER) 1989 require that information relating to health and safety at work be furnished to all employees by means of posters or leaflets in a form approved by the Health and Safety Executive.

The approved poster *"Health and Safety Law - what you should know"* should be placed in a prominent position and should contain details of the names and addresses of the enforcing authority and employment medical advisory service (EMAS). Since the modification to these Regulations (see below), the name(s) of the competent person(s) and the names and locations of trade union or other safety representatives and the groups they represent must also be included. Any change of name or address should be shown within 6 months of the alteration.

The Health and Safety Executive (HSE) may approve a particular form of poster or leaflet for use in relation to a particular industry or employment and, where any such form has been approved, the HSE shall publish it. If a poster is used, the information must be legible and up to date. The poster must be prominently located in an area to which all employees have access. If a leaflet is used, revised leaflets must be issued to employees when any similar changes occur.

MODIFICATION TO THE REGULATIONS

The Health and Safety Information for Employees (Modifications and Repeals) Regulations 1995 amended these regulations and allow the HSE to approve an alternative poster to the basic 'Health and Safety Law' poster. The basic poster required updating in order to take account of European directives and recent legal developments.

The updated poster includes two new sections which allow employers to personalise information. There is now a box for the names and location of safety representatives, and a similar one for details of competent people appointed by the employer and their health and safety responsibilities.

The earlier version of the poster could have been used until the end of June 2000. After that, the new version of the poster must be displayed and the new leaflet used.

Health and Safety (Safety Signs and Signals) Regulations (SSSR) 1996

Law considered in context / more depth in Unit 4.

Arrangement of Regulations

1) Citation and commencement.

2) Interpretation.

3) Application.

4) Provision and maintenance of safety signs.

5) Information, instruction and training.

6) Transitional provisions.

7) Enforcement.

8) Revocations and amendments.

Outline of key points

The Regulations require employers to provide specific safety signs whenever there is a risk which has not been avoided or controlled by other means, e.g. by engineering controls and safe systems of work. Where a safety sign would not help to reduce that risk, or where the sign is not significant, there is no need to provide a sign.

They require, where necessary, the use of road traffic signs within workplaces to regulate road traffic.

They also require employers to:

■ Maintain the safety signs which are provided by them.
■ Explain unfamiliar signs to their employees and tell them what they need to do when they see a safety sign.

The Regulations cover 4 main areas of signs:

1) *PROHIBITION* - circular signs, prime colours red and white, e.g. no pedestrian access.

2) *WARNING* - triangular signs, prime colours black on yellow, e.g. overhead electrics.

3) *MANDATORY* - circular signs, prime colours blue and white, e.g. safety helmets must be worn.

4) *SAFE CONDITION* - oblong/square signs, prime colours green and white, e.g. fire exit, first aid etc.

Supplementary signs provide additional information.

Supplementary signs with yellow/black or red/white diagonal stripes can be used to highlight a hazard, but must not substitute for signs as defined above.

Fire fighting, rescue equipment and emergency exit signs have to comply with a separate British Standard.

Health and Safety at Work etc. Act (HASAWA) 1974

Arrangement of Act

Preliminary

1) Preliminary.

General duties

2) General duties of employers to the employees.

3)   General duties of employers and self-employed to persons other than their employees.

4)   General duties of persons concerned with premises to persons other than their employees.

5)   [repealed].

6)   General duties of manufacturers etc. as regards articles and substances for use at work.

7)   General duties of employees at work.

8)   Duty not to interfere with or misuse things provided pursuant to certain provisions.

9)   Duty not to charge employees for things done or provided pursuant to certain specific requirements.

The Health and Safety Commission and the Health and Safety Executive

10)   Establishment of the Commission and the Executive.

11)   General functions of the Commission and the Executive.

12)   Control of the Commission by the Secretary of State.

13)   Other powers of the Commission.

14)   Power of the Commission to direct investigations and inquiries.

Health and safety regulations and approved codes of practice

15)   Health and safety regulations.

16)   Approval of codes of practice by the Commission.

17)   Use of approved codes of practice in criminal proceedings.

Enforcement

18)   Authorities responsible for enforcement of the relevant statutory provisions.

19)   Appointment of inspectors.

20)   Powers of inspectors.

21)   Improvement notices.

22)   Prohibition notices.

23)   Provisions supplementary to s.21 and 22.

24)   Appeal against improvement or prohibition notice.

25)   Power to deal with cause of imminent danger.

26)   Power of enforcing authorities to indemnify their inspectors.

Obtaining and disclosure of information

27)   Obtaining of information by the Commission, the Executive, enforcing authorities etc.

28)   Restrictions on disclosure of information.

Special provisions relating to agriculture

29-32)   [repealed].

Provisions as to offences

33)   Offences.

34)   Extension of time for bringing summary proceedings.

35)   Venue.

36)   Offences due to fault of other person.

37)   Offences by bodies corporate.

38)   Restriction on institution of proceedings in England and Wales.

39)   Prosecutions by inspectors.

40)   Onus of proving limits of what is practicable etc.

41)   Evidence.

42)   Power of court to order cause of offence to be remedied or, in certain cases, forfeiture.

Financial provision

43)   Financial provisions.

Miscellaneous and supplementary

44)   Appeals in connection with licensing provisions in the relevant statutory provisions.

45)   Default powers.

46)   Service of notices.

47)   Civil liability.

48)   Application to Crown.

49)   Adaptation of enactments to metric units or appropriate metric units.

50)   Regulations under the relevant statutory provisions.

51)   Exclusion of application to domestic employment.

52)   Meaning of work and at work.

53)   General interpretation of Part I.

54)   Application of Part I to Isles of Scilly.

Outline of key points

AIMS

1)   To protect people.

2)   To protect the public from risks which may arise from work activities.

THE MAIN PROVISIONS - SECTION 1

a)   Securing the health, safety and welfare of people at work.

b)   Protecting others against risks arising from workplace activities.

c)   Controlling the obtaining, keeping, and use of explosive and highly flammable substances.

d)   Controlling emissions into the atmosphere of noxious or offensive substances.

Duties imposed on:

a)   The employer.

b)   The self employed.

c)   Employees.

d)   Contractors and subcontractors.

e)   Designers, manufacturers, suppliers, importers and installers.

f)   Specialists - architects, surveyors, engineers, personnel managers, health and safety specialists, and many more.

EMPLOYER'S DUTIES - [TO EMPLOYEES]

Section 2(1)

To ensure, so far as *reasonably practicable*, the health, safety and welfare at work of employees.

Section 2(2)

Ensuring health, safety and welfare at work through:

- Safe plant and systems of work, e.g. provision of guards on machines.
- Safe use, handling, storage and transport of goods and materials, e.g. manual handling of boxes.
- Provision of information, instruction, training and supervision, e.g. provision of induction training.
- Safe place of work including means of access and egress, e.g. aisles kept clear.
- Safe and healthy working environment, e.g. good lighting.

Further duties are placed on the employer by:

Section 2(3)

Prepare and keep up to date a written safety policy supported by information on the organisation and arrangements for carrying out the policy. The safety policy has to be brought to the notice of employees. If there are fewer than five employees, this section does not apply.

Section 2(4)

Recognised Trade Unions have the right to appoint safety representatives to represent the employees in consultations with the employer about health and safety matters.

Section 2(6)

Employers must consult with any safety representatives appointed by recognised Trade Unions.

Section 2(7)

To establish a safety committee if requested by two or more safety representatives.

EMPLOYER'S DUTIES - [TO PERSONS NOT HIS EMPLOYEES]

Section 3

a)   Not to expose them to risk to their heath and safety e.g. contractor work barriered off.

b)   To give information about risks which may affect them e.g. location induction for contractors.

SELF EMPLOYED DUTIES

Section 3

a)   Not to expose themselves to risks to their health and safety e.g. wear personal protection.

b)   Not to expose other persons to risks to their health and safety e.g. keep shared work area tidy.

Some of the practical steps that an organisation might take in order to ensure the safety of visitors to its premises are:

- Identify visitors by signing in, badges etc.
- Provide information regarding the risks present and the site rules and procedures to be followed, particularly in emergencies.
- Provide escorts to supervise visitors throughout the site.
- Restrict access to certain areas.

PEOPLE IN CONTROL OF PREMISES

Section 4

This section places duties on anyone who has control to any extent of non-domestic premises used by people who are not their employees. The duty extends to the provision of safe premises, plant and substances, e.g. maintenance of a boiler in rented out property.

MANUFACTURERS, DESIGNERS, SUPPLIERS, IMPORTERS, INSTALLERS

Section 6

This section places specific duties on those who can ensure that articles and substances are as safe and without risks as is reasonably practicable. The section covers:

- Safe design, installation and testing of equipment (including fairground equipment).
- Safe substances tested for risks.
- Provision of information on safe use and conditions essential to health and safety.
- Research to minimise risks.

EMPLOYEES' DUTIES

Section 7

a)   To take reasonable care for themselves and others that may be affected by their acts / omissions, e.g. wear eye protection, not obstruct a fire exit.

b)   To co-operate with the employer or other to enable them to carry out their duty and/or statutory requirements, e.g. report hazards or defects in controls, attend training, provide medical samples.

Additional duties created by the Management of Health and Safety at Work Regulations employees' duties:

- Every employee shall use any equipment, material or substance provided to them in accordance with any training and instruction.
- Every employee shall inform (via supervisory staff) their employer of any (a) risk situation or (b) shortcoming in the employer's protection arrangements.

OTHER DUTIES

Section 8

No person to interfere with or misuse anything provided to secure health and safety - e.g. wedge fire door open, remove first aid equipment without authority, breach lock off systems.

Section 9

Employees cannot be charged for anything done or provided to comply with a specific legal obligation e.g. personal protective equipment, health surveillance or welfare facilities.

OFFENCES COMMITTED BY OTHER PERSONS

Section 36

- Where the commission by any person of the breach of legislation is due to the act or default of some other person, that other person shall be guilty of the offence and may be charged with and convicted of the offence whether or not proceedings are taken against the first mentioned person.
- Case law indicates that 'other person' refers to persons lower down the corporate tree than mentioned in section 37, e.g. middle managers, safety advisors, training officers; and may extend to people working on contract e.g. architects, consultants or a planning supervisor.

OFFENCES COMMITTED BY THE BODY CORPORATE

Section 37

Where there has been a breach of legislation on the part of a body corporate (limited company or local authority) and the offence can be proved to have been committed with the consent or connivance of or to be attributable to any neglect on the part of any director, manager, secretary or similar officer of the body corporate, he, as well as the body corporate, can be found guilty and punished accordingly.

ONUS OF PROOF

Section 40

In any proceedings for an offence under any of the relevant statutory provisions involving a failure to comply with a duty or requirement:

- To do something so far as is practicable.
- To do something so far as is reasonably practicable.
- It shall be for the accused to prove that the requirements were met rather than for the prosecution to prove that the requirements were not met.

Management of Health and Safety at Work Regulations (MHSWR) 1999

Law considered in context / more depth in Unit 5.

Arrangement of Regulations

1) Citation, commencement and interpretation.
2) Disapplication of these Regulations.
3) Risk assessment.
4) Principles of prevention to be applied.
5) Health and safety arrangements.
6) Health surveillance.
7) Health and safety assistance.
8) Procedures for serious and imminent danger and for danger areas.
9) Contacts with external services.
10) Information for employees.
11) Co-operation and co-ordination.
12) Persons working in host employers' or self-employed persons' undertakings.
13) Capabilities and training.
14) Employees' duties.
15) Temporary workers.
16) Risk assessment in respect of new or expectant mothers.
17) Certificate from a registered medical practitioner in respect of new or expectant mothers.
18) Notification by new or expectant mothers.
19) Protection of young persons.
20) Exemption certificates.
21) Provisions as to liability.
22) Exclusion of civil liability.
23) Extension outside Great Britain.
24) Amendment of the Health and Safety (First-Aid) Regulations 1981.
25) Amendment of the Offshore Installations and Pipeline Works (First-Aid) Regulations 1989.
26) Amendment of the Mines Miscellaneous Health and Safety Provisions Regulations 1995.

27) Amendment of the Construction (Health, Safety and Welfare) Regulations 1996.
28) Regulations to have effect as health and safety regulations.
29) Revocations and consequential amendments.
30) Transitional provision.

Schedule 1. General principles of prevention.
Schedule 2. Consequential amendments.

Outline of key points

Management of Health and Safety at Work Regulations (MHSWR) 1999 set out some broad general duties which apply to almost all kinds of work. They are aimed mainly at improving health and safety management. You may already be familiar with broad health and safety law of this kind - as it is the form taken by the Health and Safety at Work Act (HASAWA) 1974. The Regulations work in a similar way, and in fact they can be seen as a way of fleshing out what is already in the HASAWA 1974. The 1999 Regulations replace the Management of Health and Safety at Work Regulations 1992, the Management of Health and Safety at Work (Amendment) Regulations 1994, the Health and Safety (Young Persons) Regulations 1997 and Part III of the Fire Precautions (Workplace) Regulations 1997. The Principal Regulations are discussed below.

RISK ASSESSMENT (REGULATION 3)

The regulations require employers (and the self-employed) to assess the risk to the health and safety of their employees and to anyone else who may be affected by their work activity. This is necessary to ensure that the preventive and protective steps can be identified to control hazards in the workplace.

A *hazard* is defined as something with the potential to cause harm and may include machinery, substances or a work practice.

A *risk* is defined as the likelihood that a particular hazard will cause harm. Consideration must be given to the population, i.e. the number of persons who might be exposed to harm and the consequence of such exposure.

Where an employer is employing or about to employ young persons (under 18 years of age) he must carry out a risk assessment which takes particular account of:

■ The inexperience, lack of awareness of risks and immaturity of young persons.
■ The layout of the workplace and workstations.
■ Exposure to physical, biological and chemical agents.
■ Work equipment and the way in which it is handled.
■ The extent of health and safety training to be provided.
■ Risks from agents, processes and work listed in the Annex to Council Directive 94/33/EC on the protection of young people at work.

Where 5 or more employees are employed, the significant findings of risk assessments must be recorded in writing (the same threshold that is used in respect of having a written safety policy). This record must include details of any employees being identified as being especially at risk.

PRINCIPLES OF PREVENTION TO BE APPLIED (REGULATION 4)

Regulation 4 requires an employer to implement preventive and protective measures on the basis of general principles of prevention specified in Schedule 1 to the Regulations. These are:

1) Avoiding risks.
2) Evaluating the risks which cannot be avoided.
3) Combating the risks at source.
4) adapting the work to the individual, especially as regards the design of workplaces, the choice of work equipment and the choice of working and production methods, with a view, in particular, to alleviating monotonous work and work at a predetermined work-rate and to reducing their effect on health.
5) Adapting to technical progress.
6) Replacing the dangerous by the non-dangerous or the less dangerous.
7) Developing a coherent overall prevention policy which covers technology, organisation of work, working conditions, social relationships and the influence of factors relating to the working environment.
8) Giving collective protective measures priority over individual protective measures.
9) Giving appropriate instructions to employees.

HEALTH AND SAFETY ARRANGEMENTS (REGULATION 5)

Appropriate arrangements must be made for the effective planning, organisation, control, monitoring and review of preventative and protective measures (in other words, for the management of health and safety). Again, employers with five or more employees must have their arrangements in writing.

HEALTH SURVEILLANCE (REGULATION 6)

In addition to the requirements of specific regulations such as Control of Substances Hazardous to Health (COSHH) 2002 and Asbestos regulations, consideration must be given to carry out health surveillance of employees where there is a disease or adverse health condition identified in risk assessments.

HEALTH AND SAFETY ASSISTANCE (REGULATION 7)

The employer must appoint one or more competent persons to assist him in complying with the legal obligations imposed on the undertaking (including Part II of the Fire Precautions (Workplace) Regulations (FPWR) 1997). The number of persons appointed should reflect the number of employees and the type of hazards in the workplace.

If more than one competent person is appointed, then arrangements must be made for ensuring adequate co-operation between them. The Competent person(s) must be given the necessary time and resources to fulfil their functions. This will depend on the size the undertaking, the risks to which employees are exposed and the distribution of those risks throughout the undertaking.

The employer must ensure that competent person(s) who are not employees are informed of the factors known (or suspected) to affect the health and safety of anyone affected by business activities.

Competent people are defined as those who have sufficient training and experience or knowledge and other qualities to enable them to perform their functions.

Persons may be selected from among existing employees or from outside. Where there is a suitable person in the employer's employment, that person shall be appointed as the 'competent person' in preference to a non-employee.

PROCEDURES FOR SERIOUS AND IMMINENT DANGER AND FOR DANGER AREAS (REGULATION 8)

Employers are required to set up emergency procedures and appoint **competent persons** to ensure compliance with identified arrangements, to devise control strategies as appropriate and to limit access to areas of risk to ensure that only those persons with adequate health and safety knowledge and instruction are admitted.

The factors to be considered when preparing a procedure to deal with workplace emergencies such as fire, explosion, bomb scare, chemical leakage or other dangerous occurrence should include:

- The identification and training requirements of persons with specific responsibilities.
- The layout of the premises in relation to escape routes etc.
- The number of persons affected.
- Assessment of special needs (disabled persons, children etc.).
- Warning systems.
- Emergency lighting.
- Location of shut-off valves, isolation switches, hydrants etc.
- Equipment required to deal with the emergency.
- Location of assembly points.
- Communication with emergency services.
- Training and/or information to be given to employees, visitors, local residents and anyone else who might be affected.

CONTACTS WITH EXTERNAL SERVICES (REGULATION 9)

Employers must ensure that, where necessary, contacts are made with external services. This particularly applies with regard to first-aid, emergency medical care and rescue work.

INFORMATION FOR EMPLOYEES (REGULATION 10)

Employees must be provided with relevant information about hazards to their health and safety arising from risks identified by the assessments. Clear instruction must be provided concerning any preventative or protective control measures including those relating to serious and imminent danger and fire assessments. Details of any competent persons nominated to discharge specific duties in accordance with the regulations must also be communicated as should risks arising from contact with other employer's activities (see Regulation 11).

Before employing a child (a person who is not over compulsory school age) the employer must provide those with parental responsibility for the child with information on the risks that have been identified and preventative and protective measures to be taken.

CO-OPERATION AND CO-ORDINATION (REGULATION 11)

Employers who work together in a common workplace have a duty to co-operate to discharge their duties under relevant statutory provisions. They must also take all reasonable steps to inform their respective employees of risks to their health or safety which may arise out of their work. Specific arrangements must be made to ensure compliance with fire legislation (i.e. the Fire Precautions (Workplace) Regulations (FPWR) 1997).

PERSONS WORKING IN HOST EMPLOYERS' OR SELF EMPLOYED PERSONS' UNDERTAKINGS (REGULATION 12)

This regulation extends the requirements of regulation 11 to include employees working as sole occupiers of a workplace under the control of another employer. Such employees would include those working under a service of contract and employees in temporary employment businesses under the control of the first employer.

CAPABILITIES AND TRAINING (REGULATION 13)

Employers need to take into account the capabilities of their employees before entrusting tasks. This is necessary to ensure that they have adequate health and safety training and are capable enough at their jobs to avoid risk. To this end consideration must be given to recruitment including job orientation when transferring between jobs and work departments. Training must also be provided when other factors such as the introduction of new technology and new systems of work or work equipment arise.

Training must:

- Be repeated periodically where appropriate.
- Be adapted to take account of any new or changed risks to the health and safety of the employees concerned.
- Take place during working hours.

EMPLOYEES' DUTIES (REGULATION 14)

Employees are required to follow health and safety instructions by using machinery, substances, transport etc. in accordance with the instructions and training that they have received.

They must also inform their employer (and other employers) of any dangers or shortcoming in the health and safety arrangements, even if there is no risk of imminent danger.

TEMPORARY WORKERS (REGULATION 15)

Consideration is given to the special needs of temporary workers. In particular to the provision of particular health and safety information such as qualifications required to perform the task safely or any special arrangements such as the need to provide health screening.

RISKS ASSESSMENT IN RESPECT OF NEW OR EXPECTANT MOTHERS (REGULATION 16)

Where the work is of a kind which would involve risk to a new or expectant mother or her baby, then the assessment required by regulation 3 should take this into account.

If the risk cannot be avoided, then the employer should take reasonable steps to:

- Adjust the hours worked.
- Offer alternative work.
- Give paid leave for as long as is necessary.

CERTIFICATE FROM A REGISTERED MEDICAL PRACTITIONER IN RESPECT OF NEW OR EXPECTANT MOTHERS (REGULATION 17)

Where the woman is a night shift worker and has a medical certificate identifying night shift work as a risk then the employer must put her on day shift or give paid leave for as long as is necessary.

NOTIFICATION BY NEW OR EXPECTANT MOTHERS (REGULATION 18)

The employer need take no action until he is notified in writing by the woman that she is pregnant, has given birth in the last six months, or is breastfeeding.

PROTECTION OF YOUNG PERSONS (REGULATION 19)

Employers of young persons shall ensure that they are not exposed to risk as a consequence of their lack of experience, lack of awareness or lack of maturity.

No employer shall employ young people for work which:

- Is beyond his physical or psychological capacity.
- Involves exposure to agents which chronically affect human health.
- Involves harmful exposure to radiation.
- Involves a risk to health from extremes of temperature, noise or vibration.
- Involves risks which could not be reasonably foreseen by young persons.

This regulation does not prevent the employment of a young person who is no longer a child for work:

- Where it is necessary for his training.
- Where the young person will be supervised by a competent person.
- Where any risk will be reduced to the lowest level that is reasonably practicable.

(Note: Two HSE publications give guidance on the changes. HSG122 - New and expectant mothers at work: a guide for employers and HSG165 - Young people at work: a guide for employers.)

EXEMPTION CERTIFICATES (REGULATION 20)

The Secretary of State for Defence may, in the interests of national security, by a certificate in writing exempt the armed forces, any visiting force or any headquarters from certain obligations imposed by the Regulations.

PROVISIONS AS TO LIABILITY (REGULATION 21)

Employers cannot submit a defence in criminal proceedings that contravention was caused by the act or default either of an employee or the competent person appointed under Regulation 7.

EXCLUSION OF CIVIL LIABILITY (REGULATION 22)

Breach of a duty imposed by these Regulations shall not confer a right of action in any civil proceedings for those other than employees.

REVOCATIONS AND AMENDMENTS (REGULATIONS 24-29)

The Regulations:

- Revoke regulation 6 of the Health and Safety (First-Aid) Regulations (FAR) 1981 which confers power on the Health and Safety Executive to grant exemptions from those Regulations.
- Amend the Offshore Installations and Pipeline Works (First-Aid) Regulations 1989.
- Amend the Mines Miscellaneous Health and Safety Provisions Regulations 1995.
- Amend the Construction (Health, Safety and Welfare) Regulations (CHSWR) 1996.

The Regulations provide that, with some exceptions, the Fire Precautions (Workplace) Regulations (FPWR) 1997 are to be considered as health and safety regulations within the meaning of the Health and Safety at Work etc Act (HASAWA) 1974. The Regulations also make amendments to the statutory instruments as specified in Schedule 2.

TRANSITIONAL PROVISION (REGULATION 30)

The Regulations contain a transitional provision (regulation 30). The substitution of provisions in the 1999 Regulations for provisions of the Management of Health and Safety at Work Regulations (MHSWR) 1992 shall not affect the continuity of the law; and accordingly anything done under or for the purposes of such provision of the 1992 Regulations shall have effect as if done under or for the purposes of any corresponding provision of these Regulations.

Management of Health and Safety at Work and Fire Precautions (Workplace) (Amendment) Regulations 2003

Arrangement of Regulations

1) Citation and commencement.

2-6) Amendments to the Management of Health and Safety at Work Regulations 1999.

7-13) Amendments to the Fire Precautions (Workplace) Regulations 1997.

Outline of key points

AMENDMENTS TO MANAGEMENT OF HEALTH AND SAFETY AT WORK REGULATIONS 1999

2. The Management of Health and Safety at Work Regulations 1999[5] shall be amended in accordance with regulations 3 to 6 of these Regulations and any reference in those provisions to any specified provision shall, unless the context requires otherwise, be taken to be a reference to the provision so specified of the Management of Health and Safety at Work Regulations 1999.

3. For regulation 2 there shall be substituted the following regulation -

" Disapplication of these Regulations 2. -

(1) These Regulations shall not apply to or in relation to the master or crew of a ship, or to the employer of such persons, in respect of the normal ship-board activities of a ship's crew which are carried out solely by the crew under the direction of the master.

(2) Regulations 3(4), (5), 10(2) and 19 shall not apply to occasional work or short-term work involving work regarded as not being harmful, damaging or dangerous to young people in a family undertaking.

(3) In this regulation -

"normal ship-board activities" include -

    (a) the construction, reconstruction or conversion of a ship outside, but not inside, Great Britain; and

    (b) the repair of a ship save repair when carried out in dry dock; "ship" includes every description of vessel used in navigation, other than a ship belonging to Her Majesty which forms part of Her Majesty's Navy.".

4. In regulation 3(3) the words "and where" to the end shall follow and not appear in sub-paragraph (b).

5.   Regulation 19(4) shall be omitted.

6.   For regulation 22 there shall be substituted the following regulation - " Restriction of civil liability for breach of statutory duty 22. Breach of a duty imposed on an employer by these Regulations shall not confer a right of action in any civil proceedings insofar as that duty applies for the protection of persons not in his employment."

AMENDMENTS TO FIRE PRECAUTIONS (WORKPLACE) REGULATIONS 1997

7.   The Fire Precautions (Workplace) Regulations 1997[6] shall be amended in accordance with regulations 8 to 13 of these Regulations and any reference in those provisions to any specified provision shall, unless the context requires otherwise, be taken to be a reference to the provision so specified of the Fire Precautions (Workplace) Regulations 1997.

8.   In regulation 9(1) there shall be omitted the words "provisions of health and safety regulations or".

9.   In regulation 9(2) (a) (ii), for the words "premises to which" there shall be substituted the words "premises of a description specified in Part I of Schedule 1 to" and the word "apply" shall be omitted.

10.   For paragraph (ii) of regulation 9(2) (b) there shall be substituted the following paragraph -

"(ii) have effect in relation to a workplace in Great Britain other than -

    (a) an excepted workplace, or

    (b) any workplace referred to in paragraphs (i) and (ii) of paragraph (2) (a), other than a building on the surface at a mine,".

11.   After regulation 9(2) there shall be inserted the following paragraph -

" (2A) Notwithstanding that the provisions of Part II of these Regulations are not provisions forming part of the relevant statutory provisions, the provisions of Part II shall, in so far as they apply to any workplace referred to in paragraphs (i) and (ii) of paragraph (2)(a) other than a building on the surface at a mine, be deemed to be health and safety regulations for the purposes of sections 16 to 24, 26, 28, 33 to 40, 42, 46 and 47 of the 1974 Act.".

12.   At the end of regulation 9 there shall be inserted the following regulation -

"Civil liability for breach of statutory duty 9A. - (1) Subject to paragraph (2), and notwithstanding section 86 of the Fires Prevention (Metropolis) Act 1774[7], breach of a duty imposed on an employer by the workplace fire precautions legislation shall, so far as it causes damage, confer a right of action in civil proceedings.

(2) Breach of a duty imposed on an employer by the workplace fire precautions legislation shall not confer a right of action in civil proceedings insofar as that duty applies for the protection of persons not in his employment".

13.   In regulation 17 -

    (a) in paragraph (2), the words "27A (civil and other liability)" shall be omitted;

    (b) at the end of paragraph (5) there shall be inserted the following paragraph -

    "(6) Insofar as Part II of these Regulations contains any provision which is made under the 1971 Act, section 27A (a) of the 1971 Act shall not apply in respect of any contravention of such provision."

Personal Protective Equipment at Work Regulations (PPER) 1992

Arrangement of Regulations

1)   Citation and commencement.

2)   Interpretation.

3)   Disapplication of these Regulations.

4)   Provision of personal protective equipment.

5)   Compatibility of personal protective equipment.

6)   Assessment of personal protective equipment.

7)   Maintenance and replacement of personal protective equipment.

8)   Accommodation for personal protective equipment.

9)   Information, instruction and training.

10)   Use of personal protective equipment.

11)   Reporting loss or defect.

12)   Exemption certificates.

13)   Extension outside Great Britain.

14)   Modifications, repeal and revocations directive.

Schedule 1   Relevant Community.

Schedule 2   Modifications.
  Part I   Factories Act 1961.
  Part II   The Coal and Other Mines (Fire and Rescue) Order 1956.
  Part III   The Shipbuilding and Ship-Repairing Regulations 1960.
  Part IV   The Coal Mines (Respirable Dust) Regulations 1975.
  Part V   The Control of Lead at Work Regulations 1980.
  Part VI   The Ionising Radiations Regulations 1985.
  Part VII   The Control of Asbestos at Work Regulations 1987.
  Part VIII   The Control of Substances Hazardous to Health Regulations 1988.
  Part IX   The Noise at Work Regulations 1989.
  Part X   The Construction (Head Protection) Regulations 1989.
Schedule 3   Revocations.

Outline of key points

2) Personal protective equipment (PPE) means all equipment (including clothing provided for protection against adverse weather) which is intended to be worn or held by a person at work and which protects him against risks to his health or safety.

3) These Regulations do not apply to:
- Ordinary working clothes/uniforms.
- Offensive weapons.
- Portable detectors which signal risk.
- Equipment used whilst playing competitive sports.
- Equipment provided for travelling on a road.

The Regulations do not apply to situations already controlled by other Regulations i.e.
- Control of Lead at Work Regulations 1998.
- Ionising Radiation Regulations 1999.
- Control of Asbestos at Work Regulations 1987 (and as amended 1999).
- CoSHH Regulations 1999.
- Noise at Work Regulations 1989.
- Construction (Head Protection) Regulations 1989.

4) Suitable PPE must be provided when risks cannot be adequately controlled by other means. Reg. 4 shall ensure suitable PPE:
- Appropriate for the risk and conditions.
- Ergonomic requirements.
- State of health of users.
- Correctly fitting and adjustable.
- Complies with EEC directives.

5) Equipment must be compatible with any other PPE which has to be worn.

6) Before issuing PPE, the employer must carry out a risk assessment to ensure that the equipment is suitable.
- Assess risks not avoided by other means.
- Define characteristics of PPE and of the risk of the equipment itself.
- Compare characteristics of PPE to defined requirement.
- Repeat assessment when no longer valid, or significant change has taken place.

7) PPE must be maintained.
- In an efficient state.
- In efficient working order.
- In good repair.

8) Accommodation must be provided for equipment when it is not being used.

9) Information, instruction and training must be given on:
- The risks PPE will eliminate or limit.
- Why the PPE is to be used.
- How the PPE is to be used.
- How to maintain the PPE.
- Information and instruction must be comprehensible to the wearer/user.

10)   Employers shall take reasonable steps to ensure PPE is worn.
- Every employee shall use PPE that has been provided.
- Every employee shall take reasonable steps to return PPE to storage.

11)   Employees must report any loss or defect.

The Guidance on the Regulations points out:

"Whatever PPE is chosen, it should be remembered that, although some types of equipment do provide very high levels of protection, none provides 100%"

PPE includes the following when worn for health and safety reasons at work:

- Aprons.
- Adverse weather gear.
- High visibility clothing.
- Gloves.
- Safety footwear.
- Safety helmets.

- Eye protection.
- Life-jackets.
- Respirators.
- Safety harness.
- Underwater breathing gear.

Regulatory Reform (Fire Safety) Order (RRFSO) 2005

Law considered in context / more depth in Units 1, 3, 4, 5 and 6.

Introduction

At present we have to consider 5 principal pieces of legislation when considering fire safety in the workplace:

- Fire Precautions Act (FPA) 1971.
- Fire Precautions (Workplace) Regulations (FPWR) 1997.
- Management of Health and Safety at Work Regulations (MHSWR) 1999.
- Dangerous Substances & Explosive Atmosphere Regulations (DSEAR) 2002.
- Regulatory Reform Fire Safety Order (RRFSO) - due to be introduced in April 2006.

In addition to the above we would also need to consider common law.

Fire Precautions Act

This legislation is repealed when the RRFSO comes into force in October 2006.

Fire Precautions Workplace Regulations

This regulation outlines the fire safety measures that need to be achieved via the risk assessment of fire and management of fire safety within a workplace. These regulations are repealed when the RRFSO comes into force in October 2006; however in essence they have been incorporated within the RRFSO.

Management of Health and Safety at Work Regulations

It is this regulation that makes the legal requirement for fire risk assessments at present. In addition, it makes various requirements for the management of fire safety within workplaces. This regulation will continue as a stand alone health and safety regulation in the future. Again the relevant fire aspects of this regulation have been incorporated within the RRFSO.

Dangerous Substances & Explosive Atmosphere Regulations

This regulation outlines the safety and control measures that need to be taken if dangerous or flammable / explosive substances are present. This regulation will continue as a stand alone health and safety regulation in the future. Again the relevant fire aspects of this regulation have been incorporated within the RRFSO.

Regulatory Reform (Fire Safety) Order 2005 (RRFSO)

This is a new, all encompassing, fire safety order made 7th June 2005, coming into force in October 2006. As shown above it will have aspects of other legislation within it and has been compiled in such a way as to present a 'one stop shop' for fire safety legislation.

The order is split into 5 parts:

- Part 1   General.
- Part 2   Fire Safety Duties.
- Part 3   Enforcement.
- Part 4   Offences and appeals.
- Part 5   Miscellaneous.

Each part is then subdivided into the individual points or articles as they are called in the order.

Arrangement of Order

General

1)   Citation, commencement and extent.

2)   Interpretation.

3)   Meaning of "responsible person"

4)   Meaning of "general fire precautions"

5)   Duties under this Order

6)   Application to premises

7)   Disapplication of certain provisions

Part 2 - fire safety duties

8)   Duty to take general fire precautions

9)   Risk assessment

10)   Principles of prevention to be applied

11)   Fire safety arrangements

12)   Elimination or reduction of risks from dangerous substances

13)   Fire-fighting and fire detection

14)   Emergency routes and exits

15)   Procedures for serious and imminent danger and for danger areas

16)   Additional emergency measures in respect of dangerous substances

17)   Maintenance

18)   Safety assistance

19)   Provision of information to employees

20)   Provision of information to employers of self-employed from outside undertakings

21)   Training

22)   Co-operation and co-ordination

23)   General duties of employees at work

24)   Power to make regulations about fire precautions

Part 3 - enforcement

25)   Enforcing authorities

26)   Enforcement of Order

27)   Powers of inspectors

28)   Exercise on behalf of fire inspectors etc of their powers by officers of fire brigades

29)   Alterations notices

30)   Enforcement notices

31)   Prohibition notices

Part 4 - offences and appeals

32)   Offences

33)   Defence

34)   Onus of proving limits of what is practicable or reasonably practicable

35)   Appeals

36)   Determination of dispute by Secretary of State

Part 5 - miscellaneous

37)   Fire-fighters' switches for luminous tube signs etc.

38)   Maintenance of measures provided for protection of fire-fighters

39)   Civil liability for breach of statutory duty

40)   Duty not to charge employees for things done or provided

41)   Duty to consult employees

42)   Special provisions in respect of licensed etc. premises

43)   Suspension of terms and conditions of licences dealing with same matters as this Order

44) Suspension of byelaws dealing with same matters as this Order
45) Duty to consult enforcing authority before passing plans
46) Other consultation by authorities
47) Disapplication of the Health and Safety at Work etc. Act 1974 in relation to fire precautions
48) Service of notices etc.
49) Application to the Crown and to the Houses of Parliament
50) Guidance
51) Application to visiting forces etc.
52) Subordinate provisions
53) Repeals, revocations, amendments and transitional provisions

Schedule 1

Part 1   Matters to be considered in risk assessment in respect of dangerous substances
Part 2   Matters to be considered in risk assessment in respect of young persons
Part 3   Principles of prevention
Part 4   Measures to be taken in respect of dangerous substances

Schedule 2   Amendments of primary legislation

Schedule 3   Amendments of subordinate legislation

Schedule 4   Repeals

Schedule 5   Revocations

Outline of key points

PART 1 GENERAL

This part covers various issues such as the interpretation of terminology used, definition of responsible person, definition of general fire precautions, duties under the order, and its application.   The order was made on 7th June 2005, coming into force in October 2006.

The definition of general fire precautions includes:

- Measures to reduce the risk of fire on the premises and the risk of the spread of fire on the premises.
- Measures in relation to the means of escape from the premises.
- Measures for securing that, at all material times, the means of escape can be safely and effectively used.
- Measures in relation to the means for fighting fires on the premises.
- Measures in relation to the means for detecting fire on the premises and giving warning in case of fire on the premises.
- Measures in relation to the arrangements for action to be taken in the event of fire on the premises, including -
  - Measures relating to the instruction and training of employees.
  - Measures to mitigate the effects of the fire.

PART 2 FIRE SAFETY DUTIES

This part imposes a duty on the responsible person to carry out a fire risk assessment to identify what the necessary general fire precautions should be. It also outlines the principles of prevention that should be applied and the necessary arrangements for the management of fire safety. The following areas are also covered:

- Fire-fighting and fire detection.
- Emergency routes and exits.
- Procedures for serious and imminent danger and for danger areas.
- Additional emergency measures re dangerous substances.
- Maintenance.
- Safety assistance.
- Provision of information to employees, employers and self employed.
- Capabilities and training.
- Co-operation and co-ordination.
- General duties of employees.

The responsible person must appoint one or more competent persons to assist him in undertaking the preventive and protective measures required by the Order. Where the responsible person appoints persons in accordance with this requirement he must make arrangements for ensuring adequate co-operation between them. The responsible person must ensure that the number of persons appointed, the time available for them to fulfil their functions and the means at their disposal are adequate having regard to the size of the premises, the risks to which relevant persons are exposed and the distribution of those risks throughout the premises.

A competent person for the purposes of this requirement is one who has sufficient training and experience or knowledge and other qualities to enable them to provide assistance to the responsible person in undertaking the preventive and protective measures required by the Order.

PART 3 ENFORCEMENT

This part details who the enforcing authority is, which in the main is the Fire Authority, and it states they must enforce the Order. It also details the powers of inspectors and details the different types of enforcement that can be taken:

- Alterations notice.
- Enforcement notice.
- Prohibition notice.

Where served, the alteration notice requires the enforcing authority to be notified before premises are changed. The enforcement notice is similar to the improvement notices used to enforce general health and safety matters under the Health and Safety at Work etc Act (HASAWA) 1974. Similarly, the prohibition notice is similar to one under the HASAWA 1974.

PART 4 OFFENCES AND APPEALS

This part details the offences that may occur and the subsequent punishments. It also explains that the legal onus for proving that an offence was not committed is on the accused. It sets out appeals procedure for notices, which is similar to other health and safety notices except that appeals are heard in a magistrate court. A new disputes procedure is also outlined within this part.

PART 5 MISCELLANEOUS

Various matters are covered within this part the principal points being:

- 'Fire-fighters switches' for luminous tube signs etc.
- Maintenance of measures provided for the protection of fire-fighters.
- Civil liability.
- Duty not to charge employees for things done or provided.
- Duty to consult employees.
- Special provisions for licensed premises.
- Duty to consult the enforcing authority before passing plans.
- Service of notices.
- Application to crown premises.

It should be noted that if a breach of a duty imposed on an employer by or under this Order causes damage to an employee the Order confers a right for the employee to take action in civil proceedings.

The Order contains a number of schedules including ones that cover the risk assessment process, measures to be taken in respect of dangerous substances and details of the various legislation that will be repealed or amended.

Reporting of Injuries, Diseases and Dangerous Occurrences Regulations (RIDDOR) 1995

Arrangement of Regulations

1) Citation and commencement.
2) Interpretation.
3) Notification and reporting of injuries and dangerous occurrences.
4) Reporting of the death of an employee.
5) Reporting of cases of disease.
6) Reporting of gas incidents.
7) Records.
8) Additional provisions relating to mines and quarries.
9) Additional provisions relating to offshore workplaces.
10) Restrictions on the application of regulations 3, 4 and 5.
11) Defence in proceedings for an offence contravening these Regulations.
12) Extension outside Great Britain.
13) Certificates of exemption.
14) Repeal and amendment of provisions in the Regulation of Railways Act 1871, the Railway Employment (Prevention of Accidents) Act 1900 and the Transport and Works Act 1992.
15) Revocations, amendments and savings.

Schedule 1	Major Injuries.
Schedule 2	Dangerous Occurrences.
Schedule 3	Reportable Diseases.
Schedule 4	Records.
Schedule 5	Additional provisions relating to mines and quarries.
Schedule 6	Additional provisions relating to offshore workplaces.
Schedule 7	Enactments or instruments requiring the notification of events which are not required to be notified or reported under these Regulations.
Schedule 8	Revocations and amendments.

Outline of key points

The Reporting of Injuries, Diseases and Dangerous Occurrences Regulations (RIDDOR) 1995 cover the requirement to report certain categories of injury and disease sustained at work, along with specified dangerous occurrences and gas incidents, to the relevant enforcing authority. These reports are used to compile statistics to show trends and to highlight problem areas, in particular industries or companies.

REPORTING

1) When a person *dies or suffers any serious condition* specified in Schedule 1 *(Reporting of Injuries)* and Schedule 2 *(Reporting of Dangerous Occurrences)* a responsible person is to notify by the quickest possible means (usually by telephone) the enforcing authorities and must send them a written report within 10 days (F2508).

2) In cases of diseases which are linked to work activities listed in Schedule 3 *(Reporting of Diseases)* a responsible person is required to notify by the quickest possible means (usually by telephone) the enforcing authorities and must send them a written report forthwith (F2508A).

3) If personal injury results in *more than 3 days incapacity* from work off from normal duties, but does not fall in the category of "major", the written report alone is required. The day of the accident is not counted.

4) The enforcing authority is either the Health and Safety Executive or the Local Authority. The approved form for reporting is F2508 for injuries and dangerous occurrences and F2508A for diseases.

"Accident" includes:

■ An act of non-consensual physical violence done to a person at work.
■ An act of suicide which occurs on or in the course of the operation of a relevant transport system.

ROAD TRAFFIC ACCIDENTS

Road traffic accidents only have to be reported if:

■ Death or injury results from exposure to a substance being conveyed by a vehicle.
■ Death or injury results from the activities of another person engaged in the loading or unloading of an article or substance.
■ Death or injury results from the activities of another person involving work on or alongside a road.
■ Death or injury results from an accident involving a train.

NON EMPLOYEE

The responsible person must not only report non-employee deaths, but also cases that involve major injury or hospitalisation.

RECORDING

In the case of an accident at work, the following details must be recorded:

■ Date.
■ Name.
■ Nature of injury.
■ Brief description of the event.

■ Time.
■ Occupation.
■ Place of accident.

Copies of F2508 or suitable alternative records must be kept for at least 3 years. This may be held electronically provided it is printable.

DEFENCES

A person must prove that he was not aware of the event and that he had taken all reasonable steps to have such events brought to his notice.

Typical examples of major injuries, diseases and dangerous occurrences

MAJOR INJURIES (RIDDOR - SCHEDULE 1)

The list of major injuries includes:

- Any fracture, other than the finger or thumbs or toes.
- Any amputation.
- Dislocation of the shoulder, hip, knee or spine.
- Permanent or temporary loss of sight.
- Chemical, hot metal or penetrating eye injury.
- Electrical shock, electrical burn leading to unconsciousness or resuscitation or admittance to hospital for more than 24 hours.
- Loss of consciousness caused by asphyxia or exposure to a harmful substance or biological agent.
- Acute illness or loss of consciousness requiring medical attention due to any entry of substance by inhalation, ingestion or through the skin.
- Acute illness where there is a reason to believe that this resulted from exposure to a biological agent or its toxins or infected material.
- Any other injury leading to hypothermia, heat-induced illness or unconsciousness requiring resuscitation, hospitalisation greater than 24 hours.

DISEASES (RIDDOR - SCHEDULE 3)

Conditions due to physical agents and the physical demands of work

- Inflammation, ulceration or malignant disease of the skin due to ionising radiation.
- Decompression illness.
- Subcutaneous cellulitis of the hand (beat hand).
- Carpal tunnel syndrome.
- Hand-arm vibration syndrome.

Infections due to biological agents

- Anthrax.
- Hepatitis.
- Legionellosis.
- Leptospirosis.
- Tetanus.

Conditions due to chemicals and other substances

- Arsenic poisoning.
- Ethylene Oxide poisoning.
- Cancer of a bronchus or lung.
- Folliculitis.
- Acne.
- Pneumoconiosis.
- Asbestosis.
- Occupational dermatitis.

DANGEROUS OCCURRENCES (RIDDOR - SCHEDULE 2)

Dangerous occurrences are events that have the potential to cause death or serious injury and so must be reported whether anyone is injured or not. Examples of dangerous occurrences that must be reported are:

- The failure of any load bearing part of any lift, hoist, crane or derrick etc.
- The failure of any pressurised closed vessel.
- The failure of any freight container in any of its load bearing parts.
- Any unintentional incident in which plant or equipment either comes into contact with an uninsulated overhead electric line or causes an electrical discharge from such an electric line by coming into close proximity to it.
- Electrical short circuit or overload attended by fire or explosion which results in the stoppage of the plant involved for more than 24 hours.

Note: This information is a brief summary only. For full details consult HSE document L73 A Guide to RIDDOR 95.

Safety Representatives and Safety Committees Regulations (SRSC) 1977

Arrangement of Regulations

1) Citation and commencement.
2) Interpretation.
3) Appointment of safety representatives.
4) Functions of safety representatives.
5) Inspections of the workplace.
6) Inspections following notifiable accidents, occurrences and diseases.
7) Inspections of documents and provision of information.
8) Cases where safety representatives need not be employees.
9) Safety committees.
10) Power of Health and Safety Commission to grant exemption.
11) Provision as to industrial tribunals.

Outline of key points

The Safety Representatives and Safety Committees Regulations (SRSC) 1977 are concerned with the appointment by recognised trade unions of safety representatives, the functions of the representatives and the establishment of safety committees.

Representatives are appointed when a recognised trade union notifies the employer in writing. Representatives must have been employed throughout the preceding 2 years or, where this is not reasonably practicable, have had at least 2 years' experience in similar employment.

Similarly, employees cease to be representatives when:

- The employer has be notified in writing by the trade union.
- The representative ceases to be employed.
- He/she resigns.

FUNCTIONS OF TRADE UNION - APPOINTED SAFETY REPRESENTATIVES

The SRSC 1977 grant safety representatives the right to carry out certain functions as outlined below.

Functions are activities that safety representatives are permitted to carry out by legislation, but do not have a 'duty' to perform and therefore are treated as advisory actions. As a consequence the representatives cannot be held accountable for failing to carry out these activities or for the standard of the advice given, when performing their functions. They are, however, still employees and have the same consequent duties as any other employee (for example their duties under HASAWA Ss 7 and 8). Their functions as safety representatives are:

a) To take all reasonably practical steps to keep themselves informed of:
- The legal requirements relating to the health and safety of persons at work, particularly the group or groups of persons they directly represent.
- The particular hazards of the workplace and the measures deemed necessary to eliminate or minimise the risk deriving from these hazards and the health and safety policy of their employer and the organisation and arrangements for fulfilling that policy.

b) To encourage co-operation between their employer and his employees in promoting and developing essential measures to ensure the health and safety of employees, and in checking the effectiveness of these measures.

c) To carry out investigations into:
- Hazards and dangerous occurrences (incl. accidents) at the workplace.
- Complaints, by any employee he represents, relating to that employee's health, safety or welfare.

d) To carry out inspections of the workplace.

e) To bring to the employer's notice, normally in writing, any unsafe or unhealthy conditions, or unsafe working practices, or unsatisfactory arrangements for welfare at work, which comes to their attention whether during an inspection/investigation or day to day observation.

The report does not imply that all other conditions and working practices are safe and healthy or that the welfare arrangements are satisfactory in all other respects. Making a written report does not preclude the bringing of such matters to the attention of the employer or his representative by a direct oral approach in the first instance, particularly in situations where speedy remedial action is necessary. It will also be appropriate for minor matters to be the subject of direct discussion, without the need for a formal written approach.

f)   To represent the employees they were appointed to represent in consultation at the workplace with inspectors of the Health and Safety Executive and of any other enforcing authority within the Act.

g)   To receive information from inspectors in accordance with section 28(8) of the 1974 Act.

h)   To attend meetings of safety committees during which he/she attends in his capacity as a safety representative in connection with any of the above conditions.

EMPLOYERS DUTIES

The Regulations require employers to make any known information available to safety representatives which is necessary to enable them to fulfil their functions. This should include:

a)   Information about the plans and performances of the undertaking and any changes proposed, in so far as they affect the health and safety at work of their employees.

b)   Information of a technical nature about hazards to health and safety and precautions deemed necessary to eliminate or minimise them, in respect of machinery, plant, equipment, processes, systems of work and substances in use at work. This should include any relevant information provided by consultants or designers or by the manufacturer, importer or supplier of any article or substance used, or proposed to be used, at work by their employees.

c)   Information which the employer keeps relating to the occurrence of any accidents, dangerous occurrences or notifiable industrial disease and any statistical records relating to such accidents, dangerous occurrences or cases of notifiable industrial disease.

d)   Any other information specifically related to matters affecting the Health and Safety at work of his employees, including the result of any measurements taken by persons acting on his behalf in the course of checking the effectiveness of his health and safety arrangements.

e)   Information on articles or substances which an employer issues to homeworkers.

f)   Any other suitable and relevant reasonable facility to enable the representatives to carry out their functions.

TRAINING

The basis of Trades Union Congress (TUC) policy is that the union appointed safety representative will be trained on TUC approved courses. However, there is much to be gained by the employer approaching the trades unions active in his workplace with the objective of holding joint company/industry based courses. In any event it is prudent for the employer to carry out company/industry orientated training to supplement the wide industry based TUC course. The functions and training of the safety representatives should be carried out during normal working hours. The representative must receive normal earnings, this taking into consideration any bonuses which would have been earned if carrying out their normal work activities.

FUNCTIONS OF HEALTH AND SAFETY COMMITTEES

If two or more appointed safety representatives request in writing the formation of a safety committee, the employer must implement this request within three months. Consultation must take place with the representatives making the request and the appointing trade union. A basic requirement for a successful safety committee is the desire of both employee and management to show honest commitment and a positive approach to a programme of accident prevention and the establishment of a safe and healthy environment and systems of work. For any committee to operate effectively, it is necessary to determine its objectives and functions.

Objectives

a)   The promotion of safety, health and welfare at work by providing a forum for discussion and perhaps a pressure group.

b)   To promote and support normal employee/employer systems for the reporting and control of workplace problems.

Functions

a)   To review accident and occupational health trends.

b)   To review recurring problems revealed by safety audits.

c)   To consider enforcing authority reports and information releases.

d)   To consider reports on matters arising from previous safety committee meetings.

e)   To assist in the development of safety rules and systems of work and procedures.

f)   To review health and safety aspects of future development and changes in procedure.

g)   To review health and safety aspects of purchasing specifications of equipment and materials.

h)   To review renewal/maintenance programmes.

i)   To monitor safety training programmes and standards achieved.

j)   To monitor the effectiveness of safety and health communications within the workplace.

k)   To monitor the effectiveness of the Safety Policy.

This may be summarised as review and recommend on the overall direction of the health and safety programme, on specific aspects of the programme, on difficulties encountered in its implementation and to monitor the programme in both a specific and overall manner.

Composition

The membership and structure of the safety committee should be settled in consultation between management and the trade union representatives concerned. This should be aimed at keeping the total size as compact as possible, compatible with the adequate representation of the interests of management and employees. Management representatives will naturally be appointed by the management. Employee representatives will either be appointed by a recognised Trade Union (HASAWA 1974 2(4)) or, in a non-union company, elected by their colleagues. The committee suggested in HASAWA 1974 section 2 (7) will probably be the 'Company Safety Committee.' There is nothing to prevent the formation of 'works' or 'office' committees as required in order to maintain the company safety committee at a reasonable size.

Water Resources Act (WRA) 1991

Law considered in context / more depth in Unit 6.

Introduction

This Act covers protection of water against pollution and other water resource management, and only applies to England and Wales. It came into effect on 1 December 1991 and replaced corresponding sections of the Water Act 1989. It deals with some of the responsibilities of the Environment Agency.

Arrangement of Regulations

Part I   Preliminary.

Part II   Water Resources Management.

PART III   CONTROL OF POLLUTION OF WATER RESOURCES

I - Quality Objectives

82. Classification of quality of waters.

83. Water quality objectives.

84. General duties to achieve and maintain objectives etc.

II - Pollution Offences

Principal offences

85. Offences of polluting controlled waters.

86. Prohibition of certain discharges by notice or regulations.

87. Discharges into and from public sewers etc.

88. Defence to principal offences in respect of authorised discharges.

89. Other defences to principal offences.

Offences in connection with deposits and vegetation in rivers.

90. Offences in connection with deposits and vegetation in rivers.

Appeals in respect of consents under Chapter II.

91. Appeals in respect of consents under Chapter II.

III - Powers to Prevent and Control Pollution

92. Requirements to take precautions against pollution.

93. Water protection zones.

94. Nitrate sensitive areas.

95. Agreements in nitrate sensitive areas.

96. Regulations with respect to consents required by virtue of section 93 or 94.

97. Codes of good agricultural practice.

IV - Supplemental Provisions with respect to Water Pollution

98. Radioactive substances.

99. Consents required by the Authority.

100. Civil liability in respect of pollution and savings.

101. Limitation for summary offences under Part III.

Outline of key points

The Water Resources Act coupled with Part III Control of Pollution of Water Resources may need to be considered when carrying out the fire risk assessment. Under the Water Resources Act you would normally need a licence to withdraw water from a source. This is exempted if the water is required for fire fighting purposes.

As part of the 'General Fire Precautions' which have to be identified by the fire risk assessment, the 'Responsible Person' needs to 'mitigate the effects of fire on anyone in the premises and anyone in the vicinity of the premises that a fire on the premises may affect'. As a result of this statement you will need to consider the potential for environmental damage including water pollution, as this could affect people. This aspect would also cover the water run off from any fire area. This run off may need to be collected, stored or removed so as not to cause water pollution. On certain premises (especially industrial sites with chemicals) this area of concern may need additional equipment, procedures and policies in place.

Assessment

Content

> **QUESTIONS AND ANSWERS ARE REPRODUCED WITH THE KIND PERMISSION OF NEBOSH**

Written assessments - Papers A1 and FC1

At every examination a number of candidates - including some good ones - perform less well than they might because of poor examination technique. It is essential that candidates practice answering both essay-type and short answer questions and learn to budget their time according to the number of marks allocated to questions (and parts of questions) as shown on the paper.

Those of you undertaking a full course of study for the NEBOSH Certificate in Fire Safety and Risk Management will need to attend examinations for Paper A1 and Paper FC1 and complete the Practical Assessment (FC2).

Others may be undertaking a conversion course, i.e. candidates who have already achieved:

- The NEBOSH National General Certificate (Paper A1), OR
- A pass in a NEBOSH Part 1 or Part 2 Diploma, OR
- A pass in Unit A of the NEBOSH National Diploma (formerly known as Level 4).

Within the last 5 years are all exempt from Unit NGC1 of the Certificate in Fire Safety and Risk Management; therefore, such candidates **would not have to sit the A1 examination paper again** but will need to attend the examination for Paper FC1 and complete the Practical Assessment (FC2).

Each written paper (A1 and FC1) is of 2 hours duration and contains 2 sections:

- Section 1 has one question carrying 20 marks requiring quite an 'in-depth' answer. This question should be allocated 30 minutes in total. If time (e.g. 5 minutes) is given to reading, planning and checking, the time available for writing is 25 minutes. Two pages are allowed for this answer; candidates should produce approximately 1½ sides for an average answer.
- Section 2 has 10 questions each carrying 8 marks. If time (e.g. 10 minutes) is allowed for reading, planning and checking then there are 8 minutes to answer each question. One page is allowed for each of these answers, candidates should produce approximately ½ a side for an average answer.

A common fault is that candidates may fail to pay attention to the action verb in each question. The most common 'action verbs' used in Certificate examination questions are:

Define	provide a generally recognised or accepted definition.
State	a less demanding form of 'define', or where there is no generally recognised definition.
Sketch	provide a simple line drawing using labels to call attention to specific features.
Explain	give a clear account of, or reasons for.
Describe	give a word picture.
Outline	give the most important features of (less depth than either 'explain' or 'describe', but more depth than 'list').
List	provide a list without explanation.
Give	provide without explanation (used normally with the instruction 'give an example [or examples] of...').
Identify	select and name.

Other questions may start with 'what', 'when', 'how' etc. In such cases the examiners are expecting candidates to give their own explanations.

Regarding the *General* Certificate, NEBOSH questions have progressively changed to reflect practical issues that need to be managed in the workplace. Questions have increasingly reflected more than one unit of knowledge, for example, "electrical fires" which could require an understanding of Unit 10 and Unit 11 to answer adequately. This approach could similarly apply to the Certificate in Fire Safety and Risk Management, so candidates should bear this in mind when preparing for examination.

The need to understand the meaning of the 'action verb' and to read the question carefully is emphasised in the comments below that are taken from some recent examiner's reports for the *General* Certificate; (again the need to understand the question correctly will also apply to the Certificate in Fire Safety and Risk Management):

"... many answers were too brief to satisfy the requirement for an outline or description. Points made should have been supported by sufficient reasoning to show their relevance to the question."

"Some candidates, even though they identified many of the relevant factors, could not be awarded the full range of marks available because they produced a truncated list that did not properly outline the relationship between each factor and the corresponding risks."

"It was disappointing to note that some candidates again misread the question and provided outlines of the duties of employers rather than employees."

"While answers to this question were generally to a reasonable standard, many were too brief to attract all the marks that were available."

"In answering questions on Paper A2 (this will mean Paper FC1 for the NEBOSH Certificate in Fire Safety and Risk Management) practical issues should be addressed. It is not sufficient in questions such as this merely to refer to generic issues such as risk assessment and safe systems of work without providing further detail of the controls that a risk assessment might show to be necessary or the elements of a safe system of work."

"Weaker answers tended to be those that provided insufficient detail - for example, mention of "PPE" or "edge protection" should have been accompanied by some examples of what might be required and reference to the purpose that they serve."

"Some answers were extremely brief and candidates should remember that one-fifth of the marks for the entire paper are available for answers to this question (question 1). Answers are expected to be proportionate to the marks available."

NEBOSH sample questions

Element 1

1. **Identify FOUR** people or organisations who may undertake an investigation following a workplace fire, giving reasons in **EACH** case for their investigation. **(8)**

Guide to the NEBOSH Fire Safety and Risk Management Certificate, September 2005 Paper FC1 Question 2

Element 2

1. (a) **Explain**, using a suitable sketch, the significance of the 'fire triangle'. **(4)**

   (b) List **FOUR** types of ignition source that may cause a fire to occur, giving a typical workplace example of EACH type. **(4)**

March 2005, Paper A2, Question 6 (NGC)

2. **Outline** measures that should be taken to minimise the risk of fire from electrical equipment. **(8)**

March 2003, Paper A2, Question 5 (NGC)

3. (a) **Explain**, with an example, the meaning of a Class D fire. **(2)**

   (b) **List TWO** extinguishing agents suitable for use on Class D fires. **(2)**

   (c) **Outline FOUR** factors to be considered when siting portable fire fighting equipment. **(4)**

September 1996, Paper A1, Question 3 (NGC)

4. (a) **Identify FOUR** types of ignition source that may lead to a fire in the workplace. **(4)**

   (b) **Outline** ways of controlling **EACH** of the ignition sources identified in (a). **(4)**

December 1999, Paper A1, Question 8 (NGC)

Element 3

1. (a) **List FOUR** sources of ignition that should be considered when storing and using flammable solvents. **(4)**

   (b) **Identify TWO** extinguishing agents that can be used on fires that involve flammable solvents and explain their mode of action. **(6)**

   (c) **Outline** the safety precautions that should be taken when storing and using flammable solvents. **(10)**

December 1996 Paper A1 Question 1 (NGC)

2. (a) **Explain** the dangers associated with liquefied petroleum gas (LPG). **(5)**

   (b) **Describe** the precautions needed for the storage, use and transportation of LPG in cylinders on a construction site. **(15)**

June 2000 Section 1 Question 7 (NCC)

3. Prepare notes for a briefing session ('toolbox talk') associated with the prevention and control of fire on a construction site. **(40)**

December 2000 Section 1 Question 6 (NCC)

4. Timber sections are stored in an enclosed compound.

   (i) **Describe** the precautions that should be taken to prevent a fire occurring. **(4)**

   (ii) **Identify** the extinguishing media that may be used for dealing with a fire in such a situation, and **outline** how each work to extinguish the fire. **(4)**

   June 2004 Section 2 Question 4 (NCC)

5. **Outline** the precautions necessary for the safe storage and handling of small containers containing flammable solvents. **(8)**

   June 2005 Paper A2 Question 2 (NGC)

6. **Outline** the possible effect on health and safety of poor housekeeping in the workplace. **(8)**

   March 2005 Paper A1 Question 9 (NGC)

7. (a) **Explain** the health and safety benefits of restricting smoking in the workplace. **(4)**

   (b) **Outline** the ways in which an organisation could effectively implement a no-smoking police. **(4)**

   December 2001 Paper A2 Question 3 (NGC)

8. **List EIGHT** ways of reducing the risk of a fire starting in a workplace. (8)

   June 2001 Paper A1 Question 11 (NGC)

9. (a) **Explain TWO** ways in which electricity can cause a fire at work. **(2)**

   (b) **Outline** measures that should be taken to minimise the risk of fire from electrical equipment. **(6)**

   Guide to the NEBOSH Fire Safety and Risk Management Certificate, September 2005 Paper FC1 Question 3

10. **Outline** the fire risks which can arise in construction work. **(8)**

   Guide to the NEBOSH Fire Safety and Risk Management Certificate, September 2005 Paper FC1 Question 4

Element 4

1. (a) With reference to methods of heat transfer, **explain** how fire in a workplace may spread. **(8)**

   (b) **Outline** measures that should be taken to minimise the risk of fire from electrical equipment. **(8)**

   (c) **Explain** why water should not be used on fires involving electrical equipment and **identify TWO** suitable extinguishing agents that could be used in such circumstances **(4)**

   December 2004 Paper A2 Question 1 (NGC)

2. (a) With reference to the 'fire triangle', **outline TWO** methods of extinguishing fires. **(4)**

   (b) **State** the ways in which persons could be harmed by a fire in work premises. **(4)**

   March 2005 Paper A2 Question 4 (NGC)

3. **Outline** the main requirements for a safe means of escape from a building in the event of a fire. **(8)**

   December 2000 Paper A1 Question 11 (NGC)

4. (a) **Outline** the main factors to be considered in the siting of fire extinguishers. **(4)**

   (b) **Outline** suitable arrangements for the inspection and maintenance of fire extinguishers in the workplace. **(4)**

   June 2004 Paper A2 Question 2 (NGC)

5. (a) **Identify TWO** ways in which an alarm can be raised in the event of a fire in a workplace. **(2)**

   (b) **Outline** the issues to consider in the selection and siting of portable fire extinguishers. **(6)**

   September 2005 Paper A2 Question 10 (NGC)

6. **Outline** the issues that should be included in a training programme for employees on the emergency action to take in the event of fire. **(8)**

   March 1998 Paper A2 Question 9 (NGC)

7. **Identify** the **FOUR** methods of heat transfer and **explain** how **EACH** can cause the spread of fire. **(8)**

   September 2000 Paper A1 Question 9 (NGC)

8. (a) **Explain** why water should not be used as an extinguishing agent for use on fires involving:

    (i) Petroleum spirit. **(2)**

    (ii) Electrical equipment. **(2)**

    (b) **Outline** the main factors to be considered when siting portable fire fighting equipment. **(4)**

    March 1997 Paper A1 Question 8 (NGC)

9. (a) **Outline TWO** advantages and **TWO** disadvantages of using hose reels as a means of extinguishing fires. **(4)**

    (b) **Outline** the main factors to consider in the siting of hose reels. **(4)**

    September 1999 Paper A1 Question 3 (NGC)

10. **Outline** the sources of pollution which can arise in the event of a workplace fire. **(8)**

    Guide to the NEBOSH Fire Safety and Risk Management Certificate, September 2005 Paper FC1 Question 6

11. (a) **Define** the term 'Automatic Fire Detection'. **(2)**

    (b) **Outline** the factors to be considered in the selection of a fire alarm system for a workplace. **(6)**

    Guide to the NEBOSH Fire Safety and Risk Management Certificate, September 2005 Paper FC1 Question 7

12. **Outline** the measures which can be taken in the design and construction of buildings to minimise the spread of fire. **(8)**

    Guide to the NEBOSH Fire Safety and Risk Management Certificate, September 2005 Paper FC1 Question 8

Element 5

1. **Outline** the benefits of undertaking regular fire drills in the workplace. **(8)**

    December 2004 Paper A1 Question 11 (NGC)

2. **Outline** the requirements to ensure the safe evacuation of persons from a building in the event of a fire. **(8)**

    June 2003 Paper A2 Question 5 (NGC)

3. **Outline** the measures required to ensure the safe evacuation of disabled persons from a building in the event of fire. **(8)**

    Guide to the NEBOSH Fire Safety and Risk Management Certificate, September 2005 Paper FC1 Question 6

Element 6

1. **Outline** the factors to consider when carrying out a fire risk assessment of a workplace. **(8)**

    June 2000 Paper A2 Question 7 (NGC)

2. In relation to a workplace fire risk assessment, **outline** the issues that should be taken into account when accessing the *means of escape*. **(8)**

    Guide to the NEBOSH Fire Safety and Risk Management Certificate, September 2005 Paper FC1 Question 11

PLEASE REFER TO BACK OF ELEMENT FOR ANSWERS

Practical assessment - Paper FC2

The practical assessment is intended to test candidates' abilities to apply their knowledge of fire safety and risk management to a practical situation, to demonstrate their understanding of key issues and to communicate findings in an effective way. The practical assessment requires candidates to carry out unaided fire risk assessment of a workplace, followed by a summary of the fire risk assessment and then prepare a management report. It should be carried out under the control of the accredited centre and be invigilated (supervised) by a safety professional or a senior manager. The assessment must take place within 14 days of (before or after) the date of the written papers (date of the examination). Please make sure you are clear about when you will carry out the assessment and that your assessment invigilator has set the time aside.

Procedure

An observation sheet (FORM C) should be used during your fire risk assessment. There are two columns on the sheet: observations (list fire hazards, measures to prevent fire and reduce fire risk) plus actions to be taken, if any (list the immediate and longer-term actions required).

A fire risk assessment summary sheet (FORM D) and a fire risk assessment (continuation sheet) (FORM E) should also be used following your fire risk assessment. There are five columns on both FORMS D and E: significant hazards (sources of ignition and fuel), people/groups at risk, size of risk (high, medium, low), measures in place to prevent fire and/or reduce fire risks (measures to reduce fire spread, give warning and ensure safe evacuation of people) plus further actions needed to reduce risk.

The sheets must be completed during the fire risk assessment and must be included with your covering management report for marking by the assessor. Ensure that you follow the guidance given by the tutor when completing these sheets.

The maximum time allocated to the practical assessment is 2 hours and 30 minutes. This time allocation includes travel between places, i.e. the fire risk assessment observation area and the room used to complete the fire risk assessment summary sheet/s and to write the report. You should spend no more than 45 minutes making a fire risk assessment of your workplace, listing as many fire hazards/measures to prevent fire and reduce risk as possible. The remaining time should be allocated to completing the fire risk assessment summary sheet/s and writing a report to management in your own handwriting.

The whole assessment must be carried out under as near examination conditions as possible and you must not use anything previously prepared or company check lists, nor any other aids which would give you an advantage over other candidates at other centres.

You are expected to recognise physical, health and environmental fire hazards - good, as well as bad, work practices. While only short notes on each fire hazard are required, it is important that the assessor is able to subsequently identify the following:

- Where the fire hazard was located.
- The nature of the fire hazard.
- In what way, if any, the fire hazard is being controlled.
- The remedial action, where appropriate.
- Preventative action required.

During the fire risk assessment consideration should be given to other matters such as:

- Appropriateness of heating, lighting and ventilation.
- General condition of floors and gangways - particularly emergency routes.
- Cleanliness of structures.
- Mixed storage i.e. flammable, toxic oxidising materials.
- If staff are present, are they aware of the actions to take in the event of an emergency? (You are advised to spend the minimum of time on this aspect of the assessment).

On completion of the fire risk assessment and the fire risk assessment summary, you should use lined paper to produce a report, consulting your own (and only your own) notes made during the fire risk assessment. The report should be in your own handwriting. The marking sheet will be used by the assessor when marking your paper; it is not confidential and you should bear it in mind when you are writing your report.

Sample practical assessment

FIRE SAFETY AND RISK MANAGEMENT CERTIFICATE   FORM C

UNIT FC2 - THE PRACTICAL ASSESSMENT

Candidate's observation   sheet No   of

Candidate's name: _____   and number: **H** _____

Place inspected: _____   Date of inspection: _____

Observations	Actions to be taken (if any)
List fire hazards, measures to prevent fire and reduce fire risk	List all immediate and longer-term actions required
The building was a 3 storey office block with a single stairway access. The offices were generally well managed, in a good state of repair, well laid out with escape routes clear of any obstructions. There were a few issues raised which are detailed on these forms.	
Extensive use of electrical extension leads and 4 way gangs. Potential for some of extensions to be overloaded if all equipment in use at same time.	Reconfigure way equipment is plugged in (1 month). Install additional sockets (6 months)
Damage to edges of fire doors to stairwell on 1st and 2nd floor.	Install smoke seals to mitigate smoke travel into stairwell as a minimum, ideally replace doors with new ones fitted with smoke seals and intumescents. (2 months)
Large amount of old stocks of combustible materials in 1st floor storeroom. Potential to threaten escape route due to no fire detection in area.	Clean out store room and dispose of unwanted materials so as to reduce fire loading. (1 month). Consider installing smoke detection to area linked to fire alarm so as to give early warning of fire. (6 months)
Fire action notice states 'Attack fire if safe to do so', but staff asked had not been trained in use of fire equipment.	Amend wording on notices to state 'If trained' as a temporary measure. Management to discuss company policy on use of fire fighting equipment, then appoint a selection of people to be trained in practical use of equipment and policy and notices amended accordingly. (6 months).
No evidence of regular fire audits being undertaken.	Devise and implement a fire safety audit system (monthly checks recommended) Train staff to undertake audit and record findings. (3 months)

FIRE SAFETY AND RISK MANAGEMENT CERTIFICATE

FORM C

UNIT FC2 - THE PRACTICAL ASSESSMENT

Candidate's observation

sheet No   of

Candidate's name:

and number: **H**

Place inspected:

Date of inspection:

Observations List fire hazards, measures to prevent fire and reduce fire risk	Actions to be taken (if any) List all immediate and longer-term actions required
One of fire alarm call points obscured from view by temporary storage. All other call points good.	Move stored goods and maintain clear. (Immediate)
No fire detection in building	Consider installation (6 months)
No smoking policy in force in building but no evidence of illicit smoking.	Continue present management of situation. (Ongoing)
No policy in place for assisting disabled from building in the event of a fire.	No staff present with disability, but management to devise and implement system to cover visitors. (6 months)
Good secure site with access control, low potential for arson.	Continue with existing security measures and monitor local area for any arson trends. (Ongoing)
Office furniture contains polyurethane foam.	Control ignition sources and replace in future. (12 months)
Could not see many fire alarm sounders in building	Verify alarm audible throughout premises. (1 week)
No emergency lighting in building - Evening shift now worked.	Install emergency lighting (3 months)
Vents on photocopier obstructed with boxes.	Move boxes to allow air to circulate to copier. (Immediate)
Foam and CO2 extinguishers present - one of each per floor.	Keep maintained, and unobstructed as present. (Ongoing)
Walls and floors - no evidence of holes or weaknesses	Keep maintained as at present. (Ongoing)

FIRE SAFETY AND RISK MANAGEMENT CERTIFICATE
UNIT FC2 - THE PRACTICAL ASSESSMENT

FIRE RISK ASSESSMENT
SUMMARY SHEET
(FORM D)

Candidate's observation

Candidate's name: and number: **H**

Place inspected: Date of inspection:

sheet No   of

Significant hazards (sources of fuel and ignition)	People/groups at risk	Size of risk (high, medium, low)	Measures in place to prevent fire and/or reduce fire risks (measures to reduce fire spread, give warning and ensure safe evacuation of people)	Further actions needed to reduce risk
Damage to edges of fire doors to staircase on 1st and 2nd floors which could allow smoke to enter stairwell.	All persons on upper floors, but especially those on 2nd floor.	Medium	Existing doors are old style fire resistant doors fitted with self closers.	Install smoke seals to door edges to mitigate smoke travel as a minimum or replace doors with new standard fire resistant doors with smoke seals and intumescents fitted. (2 months)
Large amount of old stocks of combustible materials in 1st floor storeroom, which if ignited would cause considerable damage and has potential to threaten escape route due to lack of fire detection in building.	All persons on upper floors.	Medium	Traditional fire alarm system installed to BS 5839 with call points and some sounders but no detection. Fire resistant structure to stairwell but with damage to door edges as denoted above.	Clean out store room and dispose of unwanted materials to reduce fire loading. (1 month). Install smoke detection to area linked to fire alarm so as to give early warning. (6 months)

FIRE RISK ASSESSMENT SUMMARY (Continuation Sheet)
(FORM E)

Sheet no.   of............

Candidate's number: Date of inspection:

H

Significant hazards (sources of fuel and ignition)	People/groups at risk	Size of risk (high, medium, low)	Measures in place to prevent fire and/or reduce fire risks (measures to reduce fire spread, give warning and ensure safe evacuation of people)	Further actions needed to reduce risk
Staff not trained to use fire equipment in line with wording on fire action notice.	All staff	Low	Fire action notices sited around premises. Staff trained in evacuation process.	Management to decide on policy to be adopted, select and train staff accordingly and amend wording on notices to compliment policy. (6 months)
No evidence of regular fire audits being undertaken.	All staff	Low	Fire alarm system, fire extinguishers and emergency lighting is maintained by a third party company with records available.	Develop and implement a fire audit system with checks undertaken and recorded on a monthly basis. (3 months)
Extensive use of extension leads with 4 way gangs, potential for overating supply	All people on upper floors	Low	Electrical system installed with 'RCDs' but not circuit breakers.	Reconfigure way electrical equipment is plugged in to reduce risk of overloading supply. (1 month). Install new sockets to prevent need for extension leads. (6 months)

© ACT

FIRE RISK ASSESSMENT SUMMARY (Continuation Sheet)
(FORM E)

Candidate's number: H

Date of inspection:

Significant hazards (sources of fuel and ignition)	People/groups at risk	Size of risk (high, medium, low)	Measures in place to prevent fire and/or reduce fire risks (measures to reduce fire spread, give warning and ensure safe evacuation of people)	Further actions needed to reduce risk
An evening shift is now worked by telesales department (3rd floor) but no emergency lighting installed in building.	Staff who work in evenings	Medium	No rooms without natural or borrowed light, but insufficient light at night time.	Install emergency escape light system to 3rd floor and stairwell (3 months), consider torches as temporary measure until installed.
Fire alarm call point by exit door from 2nd floor obstructed with temporary storage of boxes.	Staff / visitors on 2nd floor.	Medium	All other call points available and not obscured. Persons escaping from 2nd floor would still be able to sound alarm by using call point by front door exit.	Move boxes from area of call point and management to ensure call points kept accessible in future by use of audit check system.. (Immediate)
No smoking policy in force in building	All persons	Low	No evidence of illicit smoking, good management system for regular breaks for staff including smokers.	Monitor situation.
Some of older office furniture on ground floor contains polyurethane foam	All persons but especially persons on ground floor	Low	Risk is only low due to good management of smoking materials and other ignition sources in area. This level of control must be maintained.	Maintain existing controls and replace furniture with combustion modified foam furniture as soon as feasible subject to budgets or within 12 months.

FIRE RISK ASSESSMENT SUMMARY (Continuation Sheet)

(FORM E)

Candidate's number: H   Date of inspection:

Significant hazards (sources of fuel and ignition)	People/groups at risk	Size of risk (high, medium, low)	Measures in place to prevent fire and/or reduce fire risks (measures to reduce fire spread, give warning and ensure safe evacuation of people)	Further actions needed to reduce risk
No policy in place for evacuation of disabled	Any person in building with a disability, plus any member of staff trying to assist them.	Low	At present no members of staff with any disability. However it is possible for a visitor with disability to enter premises upper floors via lift.	Management to devise and implement a generic policy for the evacuation of disabled. (6 months)
Access control to premises	All persons	Low	Access control gives additional security against risk of arson	Confirmation required that the access control system will fail safe and disengage on actuation of fire alarm. (1 month)
Air vents to photocopier on 1st floor obstructed with cabinet	Staff on 1st floor	Low	Photocopier well maintained and staff would become aware quickly if a fire occurred.	Re site cabinet or copier so that vents are kept clear and air movement to copier is suitable. (Immediate)
One CO2 and one water extinguisher per floor	All staff	Low	Fire fighting equipment does comply with minimum standards required under BS5306.	Consider changing water extinguishers with electrically rated spray foam extinguishers when water extinguishers become due for replacement. (Optional but recommended)

FIRE RISK ASSESSMENT SUMMARY (Continuation Sheet)
(FORM E)

Candidate's number: H

Date of inspection:

Significant hazards (sources of fuel and ignition)	People/groups at risk	Size of risk (high, medium, low)	Measures in place to prevent fire and/or reduce fire risks (measures to reduce fire spread, give warning and ensure safe evacuation of people)	Further actions needed to reduce risk
Walls and floors – no evidence of holes in structure	All staff	Low	No evidence of holes in walls / floors during inspection, but not possible to check above false ceiling.	Have ceiling voids checked to verify that there are no holes in fire structure, especially to stairs. (3 months)
No fire detection in building	All staff	Medium	Fire resistant structure to stairwell (defective doors at present)	Install fire detection throughout offices to give early warning of fire to all occupants, and by doing so guarantee exit route. (6 months)

Fire Risk Assessment Report   01/01/01

ABC Co Offices

From; ABC Smith
To: Director ABC Co Ltd

Introduction

Following on from the fire risk assessment I carried out today on the company offices I am forwarding this report to you which needs to be read in conjunction with the attached action plan form.

Summary

In general the building is in good condition with the office equipment laid out in a logical manner so that the fire escape routes are all available. The building has a fire alarm system installed, but we do not have any forms of fire detection or emergency lighting in the building. This fact coupled with some of the issues that I will report on have the potential to put the staff, any visitors, the building and the business at risk. The issues raised are as follows:-

Main Findings

1)   Fire detection

At present there is no fire detection in the building as this was acceptable when the building was built. However; we now staff and use the building in a different way to how it was used in the past, plus legislative changes force us to rethink policies. Considering the points raised in this report, I have a concern that we could have a fire start on a lower floor when the floor is unoccupied and staff on upper floors (especially the tele-sales team) would not be aware of the fire until it is too late and their only exit route is cut off. This scenario in fact breeches legislation the Regulatory Reform (Fire Safety) Order 2005 and as such we should install fire detection to the building.

2)   Defective fire doors

The fire doors between the stairwell on the 1st and 2nd floors and the office areas are damaged. The damage is on the edges of the doors and would have the effect that if there were to be a fire on either floor smoke would probably leak past the doors into the stairwell. As you know we only have the one stairwell, so it is critical that it remains useable to all of us for escape in the event of a fire. You could get a company in to repair the doors, but I would suggest that this would only be a short term fix and as such false economy. It would be better to replace the doors with new ones which will have the modern smoke seals fitted to them. In this way, should fire happen the smoke will be prevented from entering the stairwell and so give us a guaranteed escape route. The doors on the first floor are even more important than the doors on the second floor so we need to look at these doors first. The reason for this is that the first floor is often unoccupied when the sales team are out. As we do not have any fire detection in the building the fire could happen on the first floor undetected. The present defects breeches legislation : the Regulatory Reform (Fire Safety) Order 2005.

3)   Lack of emergency lighting

The company has recently started an evening shift with the telesales staff on the 3rd floor. If a fire happens which effects the electrical circuits to the lighting, these staff will not be able to escape from the building safely. This breeches legislation : the Regulatory Reform (Fire Safety) Order 2005. It is vital therefore that we install an emergency lighting system to BS5266 standard in at least the 3rd floor and the stairwell.

4)   Storage of combustibles

The storeroom on the 1st floor is full of combustibles (old books, papers, files). Most of this looks like very old information that we probably do not need to keep. At present it is causing an excess fire loading on this floor, when coupled with the fact that this floor is often unoccupied it poses a risk to people on the upper floors due to the lack of fire detection. This store must be cleared out and any remaining combustibles ideally stored in metal cabinets.

Whilst the above are the most critical flaws there are a number of items requiring attention and reference must be made to the fire risk assessment summary sheets and observation sheets that are attached.

Conclusion

Fire safety standards appear to be acceptable at face value, but as you can see from this report there are various failings which must be addressed as a matter of urgency to lower the risks to an acceptable level for the people in the building, the building itself and the business. The general day to day management of the building and the staff awareness of the actions to take in the event of a fire are very positive. We need to improve the few areas of weakness and build on the positives to ensure protection from fire for the company.

NEBOSH sample questions - answers

Element 1

1. **Identify FOUR** people or organisations who may undertake an investigation following a workplace fire, giving reasons in **EACH** case for their investigation. **(8)**

    Guide to the NEBOSH Fire Safety and Risk Management Certificate, September 2005 Paper FC1 Question 2

Fire & Rescue Service Officer

A fire officer will investigate fires to establish the cause of the fire. This information is required so that it can be fed into national statistics, which hopefully can then assist in the prevention of future fires.

Police Officer

A police officer will investigate a fire incident (probably in partnership with the fire officer) if it is believed that the cause of fire is arson, especially if an injury or death has occurred. The police officer will be especially interested in the gathering of evidence to be used in any future court case.

Insurance Investigator

An insurance investigator may undertake an investigation of a fire as a way of verifying the validity or otherwise of an insurance claim that has (or will be) made.

Fire Consultants

Fire consultants may be asked to undertake an investigation of a fire by the owner / occupier. This could for example be to ascertain where a system failure occurred which led to the fire being caused and from this the company could hopefully prevent such an incident re-occurring.

Element 2

1. (a) **Explain**, using a suitable sketch, the significance of the 'fire triangle'. **(4)**

    (b) List **FOUR** types of ignition source that may cause a fire to occur, giving a typical workplace example of EACH type. **(4)**

    March 2005, Paper A2, Question 6 (NGC)

For part (a), you should be able to outline two methods of extinguishing fires by choosing from: starvation (by removing the fuel); smothering (by eliminating the oxygen supply with the use of foam or carbon dioxide); cooling (with water); and interfering chemically with the combustion process (by using a dry powder extinguisher).

In a fire situation, people may be harmed by being burned, by inhaling toxic fumes, by the effects of smoke and by a depleted oxygen supply. There is also the possibility of being injured by a structural failure of a building, or by being crushed or suffering some other type of injury in attempting to escape.

2. **Outline** measures that should be taken to minimise the risk of fire from electrical equipment.

    March 2003, Paper A2, Question 5 (NGC)

Answers include: the proper selection of equipment to ensure its suitability for the task, pre-use inspection by the user, establishing correct fuse ratings, ensuring circuits and sockets are not overloaded, disconnecting or isolating the equipment when it is not in use, and ensuring that electric motors do not overheat (e.g. by checking that vents are uncovered). Additional measures include the need to uncoil cables {particularly extension leads} to prevent the build up of heat and protecting cables from mechanical damage. Importantly, electrical equipment and systems should be subject to regular inspection, testing and maintenance by competent persons. This should ensure, for instance, that contacts are sound, thereby reducing the likelihood of electrical arcing.

3. (a) **Explain**, with an example, the meaning of a Class D fire. **(2)**

    (b) **List TWO** extinguishing agents suitable for use on Class D fires. **(2)**

    (c) **Outline FOUR** factors to be considered when siting portable fire fighting equipment. **(4)**

    September 1996, Paper A1, Question 3 (NGC)

For part (a), reference to a Class D fire being one involving metals such as sodium, magnesium or aluminium.

For part (b), two agents from:
- *sand*
- *dry powder*
- *powdered graphite*
- *powdered limestone*
- *soda ash*

For part (c), reference to four factors from:
- *proximity to the fire hazard*

- ■ proximity to exits / escape routes
- ■ protection against the environment
- ■ provision of suitable fixtures and fittings
- ■ accessibility
- ■ visibility
- ■ in multi-storey buildings, the same position on each floor

4. (a) **Identify FOUR** types of ignition source that may lead to a fire in the workplace. **(4)**

   (b) **Outline** ways of controlling **EACH** of the ignition sources identified in (a). **(4)**

December 1999, Paper A1, Question 8 (NGC)

For part (a), the more common examples of ignition sources to be identified are smoking materials, hot work, electricity and reactive chemicals.

For smoking materials, reference should be made to smoking bans or restrictions, and the provision of designated smoking areas; for electricity, reference should be made to the means of preventing excess current and arcing (e.g. correctly rated fuses, sound connections etc) together with routine inspection, testing and maintenance; for hot work, appropriate control measures would involve the use of a pemit-to-work system, minimising the amount of combustible matter and routine checks for smouldering material on completion of the work; and, for chemicals, reference should be made to the need for appropriate storage facilities and the separation of incompatible chemicals.

Element 3

1. (a) **List FOUR** sources of ignition that should be considered when storing and using flammable solvents. **(4)**

   (b) **Identify TWO** extinguishing agents that can be used on fires that involve flammable solvents and explain their mode of action. **(6)**

   (c) **Outline** the safety precautions that should be taken when storing and using flammable solvents. **(10)**

December 1996 Paper A1 Question 1 (NGC)

For part (a), reference to four sources from:

- ■ naked flames
- ■ hot surfaces
- ■ hot work
- ■ friction
- ■ electrical sparks
- ■ specific examples from the above categories

For part (b), reference to two agents from:

- ■ dry powder
- ■ foam
- ■ inert gas
- ■ sand
- ■ with an appropriate explanation of their mode of action (e.g. cooling effects / the exclusion of oxygen).

In part (c), the terms '**safety** precautions' and '**flammable** solvents' were intended to maintain the focus on fire rather than the health aspects of hazardous substances. Reference was needed to:

- ■ quantities stored
- ■ container types
- ■ storage facilities (storerooms / cabinets)
- ■ exclusion of ignition sources
- ■ signs and markings
- ■ prevention of vapour build-up
- ■ training
- ■ provision of fire-fighting equipment
- ■ emergency arrangements

2. (a) **Explain** the dangers associated with liquefied petroleum gas (LPG). **(5)**

   (b) **Describe** the precautions needed for the storage, use and transportation of LPG in cylinders on a construction site. **(15)**

June 2000 Section 1 Question 7 (NCC)

For part (a), the dangers of liquefied petroleum gases relate to their flammable and explosive properties and to the fact that they are stored under great pressure, and hence low temperatures, to retain their liquid state. On release, liquefied petroleum gases revert to their gaseous state where the gas is at least 250 times greater in volume than when stored as a liquid. The fact that the gas is heavier than air is of great significance to those, for instance, working in excavations. The asphyxiating properties of LPG were mentioned by many in this context. Reference to correct manual handling issues when LPG is stored in cylinders, could also be included.

Better answers to part (b) are those that addressed storage, use and transportation issues in order - as the question suggested.

The key points to be made are that cylinders should not be stored inside buildings but normally in a meshed cage in the open air *protected from direct sunlight and with appropriate warning signs. Cylinders should be stored in an upright position, away from oxygen, oxidising agents and flammable materials, and full cylinders should be stored separately from empty ones. A foam or powder fire extinguisher should always be available.*

When in use, cylinders should be fixed securely in an upright position and fitted with gauges and flash-back arrestors. All connections should be good and tested for leaks. Other precautions may depend on the particular circumstances and the use to which the LPG is put but hot work permits, ventilation at top and bottom of rooms, and fire checks after use are all relevant issues. In all cases, fire extinguishers must be available.

On site, cylinders should be transported on purpose-made trolleys. Transportation in vehicles should only be carried out if the vehicle is suitable (for example not in an enclosed van) and if the cylinders can be kept upright and secured against movement. Vehicles should be equipped with a fire extinguisher and drivers provided with training and personal protective equipment such as gloves.

3.   Prepare notes for a briefing session ('toolbox talk') associated with the prevention and control of fire on a construction site. **(40)**

December 2000 Section 1 Question 6 (NCC)

A toolbox talk is an effective method for giving health and safety information to employees on a construction site. The identification of hazards, such as flammable liquids, timber and waste materials, together with possible sources of ignition, such as smoking materials and hot work, would have produced a good starting point. In identifying prevention techniques, good housekeeping, hot work permits, restriction of smoking, the safe storage of flammable and combustible materials, and ensuring electrical safety could be included in answers.

4.   Timber sections are stored in an enclosed compound.

   (i)   **Describe** the precautions that should be taken to prevent a fire occurring. **(4)**

   (ii)   **Identify** the extinguishing media that may be used for dealing with a fire in such a situation, and **outline** how each work to extinguish the fire. **(4)**

June 2004 Section 2 Question 4 (NCC)

For Part (i) precautions could include: separating the stored timber from flammable materials, keeping the area clear of paper and other rubbish and vegetation, and prohibiting smoking or operations involving hot work near to the compound.

For part (ii), reference to the use of water to cool the fire and dampen adjacent timber, and foam to exclude oxygen. Others could include the use of sand or carbon dioxide.

5.   **Outline** the precautions necessary for the safe storage and handling of small containers containing flammable solvents. **(8)**

June 2005 Paper A2 Question 2 (NGC)

Precautions that should be considered when storing and handling flammable solvents in small containers include selecting containers that are suitable for the purpose; labeling the containers clearly with information about their contents; marking the storage area in which they are held; ensuring that empty containers are tightly closed and stored outside the building or in a store constructed of fire resisting materials; taking measures to prevent vapour build-up and to prevent or reduce the impact of spillages by using non-spill caps or bunding the area where the containers are held; removing likely sources of ignition and limiting the quantities stored and the amounts in use.

6.   **Outline** the possible effect on health and safety of poor housekeeping in the workplace. **(8)**

March 2005 Paper A1 Question 9 (NGC)

Slips, trips, falls, falling objects and fires are well-known effects associated with poor housekeeping. There may also be an increased risk of ill-health where hazardous substances are involved or where rubbish attracts rodents and other pests. Specific reference could have been made to blocked fire exits and internal transport issues. Suggestion that poor storage of materials can lead to musculo-skeletal problems and that a generally untidy workplace introduces the possibility of increased levels of stress in the workforce, leading to a lowering of morale which, in itself, can result in a reluctance to adhere to accepted health and safety procedures.

7.   (a)   **Explain** the health and safety benefits of restricting smoking in the workplace. **(4)**

   (b)   **Outline** the ways in which an organisation could effectively implement a no-smoking police. **(4)**

December 2001 Paper A2 Question 3 (NGC)

For part (a):
- *as a reduction in the risk of fire,*
- *an improvement in general cleanliness and*
- *a reduction in the exposure of non-smoking staff to cigarette smoke, which can have an irritant effect as well as causing possible long term health damage.*

Other reasons include the promotion of health generally and the avoidance of conflict between smokers and non-smokers.

For part (b) the initial point to be made was that the policy should be clear in its intent and communicated to all staff. This might be achieved by the use of notice boards, leaflets and other forms of propaganda, while there would also need to be consultation with employees to encourage ownership. Management's part in the process could be helpful by setting an example and also provide help to employees in the form of counselling. Finally, the provision of designated smoking areas and the use of disciplinary procedures are suggested as options.

8. **List EIGHT** ways of reducing the risk of a fire starting in a workplace. **(8)**

June 2001 Paper A1 Question 11 (NGC)

Answers could include:

- *the control of smoking and smoking materials,*
- *good housekeeping to prevent the accumulation of waste paper and other combustible materials,*
- *regular lubrication of machinery,*
- *frequent inspection of electrical equipment for damage,*
- *ensuring ventilation outlets on equipment are not obstructed,*
- *controlling hot work,*
- *the provision of proper storage facilities for flammable liquids and the segregation of incompatible chemicals.*

9. (a) **Explain TWO** ways in which electricity can cause a fire at work. **(2)**

   (b) **Outline** measures that should be taken to minimise the risk of fire from electrical equipment. **(6)**

Guide to the NEBOSH Fire Safety and Risk Management Certificate, September 2005 Paper FC1 Question 3

Electrical equipment generates heat or produces sparks. This equipment should not be placed where it could lead to the uncontrolled ignition of any substance. This can result from wiring defects, overheating, poor electrical connections or incorrect fuse rating.

The equipment may itself explode or arc violently, and act as a source of ignition of flammable vapours, gases, liquids or dust through electric sparks, arcs or high surface temperatures.

Equipment must be suitable for the power demanded and conditions in which it is to be used. The environmental conditions in which equipment is used has significant impact on selection of equipment. Equipment used for outdoor work will need protection against the ingress of water. Special equipment such as intrinsically safe lighting should be used in flammable or explosive atmospheres. Dust in the atmosphere increases the risk of fire/explosion.

Equipment must be used in such a manner to prevent damage from occurring. Construction. Inspection, testing and maintenance should be in place to identify, examine and remove deficiencies in equipment such as

- *Wiring defects such as insulation failure due to age / abrasion / heat/ damp/ chemical attack;*
- *Overheating of cables or other electrical equipment through overloading with currents above their design capacity;*
- *Incorrect fuse rating;*
- *Inadequate connections, where cables are not secured by the outer sheath on entry to the equipment or where individual conductor insulation / the conductor wire itself is visible.*

10. **Outline** the fire risks which can arise in construction work. **(8)**

Guide to the NEBOSH Fire Safety and Risk Management Certificate, September 2005 Paper FC1 Question 4

- *Increased sources of ignition / flammable substances and materials*
- *Accumulation and storage of flammable building materials or wastes on site*
- *Introduction of additional electrical equipment, or other sources of ignition*
- *Use of hot work processes such as welding, grinding or cutting, raising sparks*
- *Introduction of flammable products - adhesives, solvents, fuels, gases to a site*
- *Contact with live electrical or gas services during work.*
- *Possibility of uncontrolled heat sources/fuels introduced by contractors*

Element 4

1. (a) With reference to methods of heat transfer, **explain** how fire in a workplace may spread. **(8)**

   (b) **Outline** measures that should be taken to minimise the risk of fire from electrical equipment. **(8)**

   (c) **Explain** why water should not be used on fires involving electrical equipment and **identify TWO** suitable extinguishing agents that could be used in such circumstances **(4)**

December 2004 Paper A2 Question 1 (NGC)

To explain that heat can be transferred through metal beams or other parts of a structure by conduction; it can be carried by rising air currents (convection) to cause a build-up of hot gases under ceilings; it can be transferred through the air by radiation causing heating of material at a distance from a fire; and, perhaps what should have been the most obvious, combustible material in direct contact with flames can itself catch fire. The purist might argue that the last of these, direct burning, is simply a combination of the other three main methods but, in fire safety terms, it is normally treated as a method in its own right.

For part (b), measures that could be outlined include: ensuring the suitability of the chosen equipment for the task (eg compliance with recognised standards, intrinsically safe equipment in flammable atmospheres, etc); circuit overload prevention, for instance by avoiding the use of multi-way adapters in single sockets; the use of correctly rated fuses and thermal cut-outs; switching off or isolating equipment when not in use; ensuring that vents remain uncovered; uncoiling cables and extension leads; pre-use inspection of equipment for visible damage to cables, plugs or connectors; and a programme of formal inspection and testing of equipment and electrical systems by a competent person.

For part (c), using water on an electrical fire can lead to electric shock since water is a good conductor of electricity. Carbon dioxide and dry powder were correctly identified as suitable extinguishing agents where electrical equipment is involved but references to halon were discounted due to the fact that its use has been banned.

2. (a)   With reference to the 'fire triangle', **outline TWO** methods of extinguishing fires. **(4)**

    (b)   **State** the ways in which persons could be harmed by a fire in work premises. **(4)**

March 2005 Paper A2 Question 4 (NGC)

For part (a), the two methods of extinguishing fires by choosing from: starvation (by removing the fuel); smothering (by eliminating the oxygen supply with the use of foam or carbon dioxide); cooling (with water); and interfering chemically with the combustion process (by using a dry powder extinguisher).

For part (b) in a fire situation, people may be harmed by being burned, by inhaling toxic fumes, by the effects of smoke and bya depleted oxygen supply. Reference could also be included to the possibility of being injured by a structural failure of a building, or by being crushed or suffering some other type of injury in attempting to escape.

3. **Outline** the main requirements for a safe means of escape from a building in the event of a fire. **(8)**

December 2000 Paper A1 Question 11 (NGC)

The term 'means of escape' refers to methods and facilities available to enable someone to escape from a fire within a building and move to a place of safety. Issues include:

- *the need to ensure that there are at least two escape routes in different directions, neither of which should exceed the recognised travel distance;*
- *the fire integrity of the actual escape route;*
- *the ability of fire doors to open easily in an outward direction and to self-close;*
- *the importance of ensuring that escape routes are clearly marked, provided with emergency lighting and kept free from obstruction; and*
- *the provision of a safe place clearly signed as an assembly point.*

4. (a)   **Outline** the main factors to be considered in the siting of fire extinguishers. **(4)**

    (b)   **Outline** suitable arrangements for the inspection and maintenance of fire extinguishers in the workplace. **(4)**

June 2004 Paper A2 Question 2 (NGC)

For part (a), typical factors include accessibility, visibility, proximity to exits and escape routes, travel distances from the extinguisher to the possible location of a fire, the means of supporting the equipment off the ground and without causing obstruction, and the need to protect extinguishers from the weather and other sources of damage.

Part (b) Good answers clearly differentiate between the respective roles of inspection and maintenance. Frequent visual inspection of fire extinguishers is required to ensure that they are in place, have not lost pressure and bear no obvious damage. More thorough, less frequent inspections should check in addition that safety clips are functioning and that there are no signs of corrosion. Maintenance, on the other hand, is something rather more extensive and usually involves an annual examination and test by a competent person according to the manufacturer's instructions in order to ensure the integrity of the extinguisher, with the date of the examination recorded on the extinguisher and the replacement of equipment found to be faulty. At suitable intervals (eg five-yearly), an extinguisher should be discharged and refilled. At greater intervals, perhaps every twenty years, an extinguisher should undergo complete overhaul or replacement.

5. (a)   **Identify TWO** ways in which an alarm can be raised in the event of a fire in a workplace. **(2)**

    (b)   **Outline** the issues to consider in the selection and siting of portable fire extinguishers. **(6)**

September 2005 Paper A2 Question 10 (NGC)

For part (a), two ways of raising the alarm in the event of fire, including, automatic methods such as smoke detectors and manually operated devices such as break glass alarms.

Part (b), for selection the relevant criteria would be the type of extinguisher required relative to the fire class and the number to be provided relative to the size of the premises.

6. **Outline** the issues that should be included in a training programme for employees on the emergency action to take in the event of fire. **(8)**

March 1998 Paper A2 Question 9 (NGC)

Reference to:

- *recognition of fire alarms and knowledge of the action to be taken*
- *meanings of emergency signs*
- *location of fire escape routes and assembly points*
- *requirements for safe evacuation (eg non-use of lifts, no running, etc)*
- *location and operation of call points and other means of raising the alarm*
- *location and use of fire-fighting equipment*
- *consideration of people with special needs*
- *the identity and role of fire marshals*

7.   **Identify** the **FOUR** methods of heat transfer and **explain** how **EACH** can cause the spread of fire. **(8)**

September 2000 Paper A1 Question 9 (NGC)

The four methods of heat transfer that can be identified in answer to this question are conduction, convection, radiation and direct burning. Heat can be transferred through metal beams or other parts of a structure by conduction; it can be carried by rising air currents (convection) to cause a build-up of hot gases under ceilings; it can be radiated through the air causing heating of material at a distance from a fire; and, perhaps what should have been the most obvious, combustible material in direct contact with flames can itself catch fire.

8.   (a)   **Explain** why water should not be used as an extinguishing agent for use on fires involving:

       (i)   Petroleum spirit. **(2)**

       (ii)   Electrical equipment. **(2)**

      (b)   **Outline** the main factors to be considered when siting portable fire fighting equipment. **(4)**

March 1997 Paper A1 Question 8 (NGC)

For part (a)(i), reference to:
■   *the fact that petroleum spirit floats on water*
■   *the fact that it will therefore either continue to burn or vaporise to form an explosive cloud*
■   *the fact that water may spread the fire over a greater area*

For part (a)(i), reference to:
■   *the electrical conductivity of water*
■   *the danger of short circuits leading to the risk of shock and explosion*

For part (b), reference to:
■   *accessibility*
■   *proximity to exits and escape routes*
■   *visibility and signage*
■   *location off the ground with adequate support*
■   *suitable type in relation to the hazards present*
■   *protection against damage / weather*

9.   (a)   **Outline TWO** advantages and **TWO** disadvantages of using hose reels as a means of extinguishing fires. **(4)**

      (b)   **Outline** the main factors to consider in the siting of hose reels. **(4)**

September 1999 Paper A1 Question 3 (NGC)

Part (a) the advantages of a continuous supply of water that is at a greater force than that from an extinguisher, with the benefit that users do not need to place themselves as close to the fire. The disadvantages relate to the effort required to position and operate the hose, the possibility of the hose compromising fire and smoke barriers by passing through doorways and wedging doors in the open position, and the fact that water is not a suitable extinguishing agent for some types of fires.

The factors to be considered in the siting of hose reels, for part (b), relate to issues of accessibility, utility and avoiding obstruction. Comprehensive answers should refer to siting hose reels adjacent to an exit or at the top of a staircase (so that the user is placed between the fire and an escape route), in particular positions where they can be extended to reach all parts of the building, and, wherever possible, in recesses to prevent obstruction of an escape route. An additional factor that may need to be considered is the siting of hose reels in positions where they are less prone to vandalism and misuse.

10.   **Outline** the sources of pollution which can arise in the event of a workplace fire. **(8)**

Guide to the NEBOSH Fire Safety and Risk Management Certificate, September 2005 Paper FC1 Question 6

Smoke / Toxic fumes
■   *Fires give off large quantities of smoke and toxic fumes which may contain other pollutant constituents depending on the fuel source.*
■   *Smoke*
■   *Smoke consists of small particles or partially burnt, carbonaceous materials. The colour, size and quantity of these particles will determine the thickness of the smoke; water vapour also thickens smoke.*
■   *Smoke and its by products are normally very corrosive and as such can cause long term damage to buildings and materials unless cleaned correctly following a fire.*

Toxic / Corrosive Gases
■   *Dependent on their individual chemical composition, various polymers and plastics release toxic gaes when burnt. These include carbon monoxide, phenol and phenolic compounds, nitrogen dioxide, hydrogen cyanide and various organic nitriles (cyanides), all of which are toxic.*
■   *Other substances, such as PVC, release toxic chlorinated and other halogenated compounds in the fire gases. These in combination with water form acidic solutions which are corrosive, attacking metals and concrete. This may lead to structural failure with accompanying uncontrolled release of substances.*

Run-off of contaminated fire-fighting water/foam
■   *Any water used when fighting a fire that is not turned into steam with the heat from the fire will run off the fire as polluted water, making solutions of many of the compounds listed above.*

- *The pollutants may escape from the site into the water environment by a number of pathways. These include:-*
- - *the site's surface water drainage system, either directly or via off-site surface water sewers.*
- - *direct run-off into nearby watercourses or onto ground, with potential risk to groundwaters.*
- - *via the foul drainage system, with pollutants either passing unaltered through a sewage treatment works or affecting the performance of the works, resulting in further environmental damage.*
- - *Solid waste*
- *Residual partially combusted, smoke damaged or heat affected materials will need to be disposed of as waste.*

11. (a) **Define** the term 'Automatic Fire Detection'. **(2)**

    (b) **Outline** the factors to be considered in the selection of a fire alarm system for a workplace. **(6)**

    Guide to the NEBOSH Fire Safety and Risk Management Certificate, September 2005 Paper FC1 Question 7

Automatic Fire Detection systems installed in buildings, which identify fire fault conditions such as heat, smoke or radiation without the interaction of people.

Types of fire alarm include:

- **Voice:** *Simplest and most effective but very limited size of workplaces,*
- **Manual alarms:** *Rotary gong, hand bell etc. again limited in scale of buildings,*
- **Call points with sounders:** *Standard system operation of one call point sounds alarm throughout workplace*
- **Automatic system:** *Standard system operation of one call point sounds alarm throughout workplace with automatic fire detection*

Factors affecting selection of an alarm:

Occupants

- *Number of occupants - need for total evacuation or priority evacuation of areas of high risk first*
- *Vulnerability of occupants - restricted awareness of alarm condition through old age or disability*

Type of Workplace

- *Layout of workplace - design and construction of building and structural fire precautions in place*

Use of workplace

- *activities conducted at the workplace, including any interaction with the public, and the use/storage of flammable materials*

Vulnerability of surrounding area

- *degree of protection required, COMAH status.*

12. **Outline** the measures which can be taken in the design and construction of buildings to minimise the spread of fire. **(8)**

    Guide to the NEBOSH Fire Safety and Risk Management Certificate, September 2005 Paper FC1 Question 8

Selection of Material

- *Materials have varying fire resistance. Appropriate selection should be made according to required tolerances*
- *Brick / blockwork perform well in fires. Dependant upon the materials, workmanship, thickness, and the load carried, fire resistance of 30 mins to 2 hours may be achieved.*
- *Steel and other metals are extensively used but can be affected by fire at relatively low temperatures and may increase the risk of fire spread by radiating heat to other parts of the building. Therefore, they must be protected by fire retardant materials, such as encasing in concrete; fire retardant boards or spray coatings.*
- *Timber performs very well in fires as long as it is of sufficient size that as its outer coat burns away, there is still sufficient strength to do its task, such as oak beams. Generally timber does not fail rapidly in a fire unlike steel.*
- *Glass generally performs poorly in a fire unless it is fire resistant glass. At high temperatures glass will melt and sag, which is why the traditional fire resistant glass has wire within it.*

Openings and voids

- *Consideration should be given to the protection of openings and voids by the use of fire barriers such as fire shutters, cavity barriers and fire curtains.*
- *Stairwells lift shafts and ceiling voids can create a chimney effect*

Compartmentisation rather than open plan offices

Controlling the storage of combustible materials

Element 5

1. **Outline** the benefits of undertaking regular fire drills in the workplace. **(8)**

    December 2004 Paper A1 Question 11 (NGC)

It is generally accepted that fire drills form an important part of ensuring the safety of employees at a workplace. Answers that can be included: satisfying a legal requirement, or one specified in a fire certificate, to provide instruction to employees on the actions to be taken in emergency situations; checking that the alarm can be heard in all parts of the premises; testing the effectiveness of the evacuation procedures both generally and in relation to specific requirements (such as the need to ensure the safety of disabled employees and visitors); familiarising employees (particularly those new to the undertaking) with the alarms, evacuation procedures, escape routes and assembly points so that, in the case of a real emergency, they would know the actions to take; and providing an opportunity for fire wardens and others with specific functions to practise their designated roles.

2. **Outline** the requirements to ensure the safe evacuation of persons from a building in the event of a fire. **(8)**

June 2003 Paper A2 Question 5 (NGC)

The requirements to ensure the safe evacuation of persons from a building in the event of a fire could be outlined as:

- *the means for raising the alarm.*
- *an acceptable distance to the nearest available exit.*
- *escape routes of sufficient width.*
- *clear signing of escape routes.*
- *the provision of emergency lighting.*
- *escape routes kept clear of obstructions with fire doors closed to prevent the spread of smoke.*
- *the provision of fire-fighting equipment.*
- *the appointment of fire marshals.*
- *procedures for the evacuation of those with a physical impairment (in relation to hearing, sight or mobility).*
- *the need to practise the evacuation plan at regular intervals.*
- *good evacuation.*

3. **Outline** the measures required to ensure the safe evacuation of disabled persons from a building in the event of fire. **(8)**

Guide to the NEBOSH Fire Safety and Risk Management Certificate, September 2005 Paper FC1 Question 5

- *If member of staff or regular visitor talk to individual and ascertain type and level of disability then design, agree, record and practice the procedure for the individual's safe escape in the event of a fire (PEEP – Personal Emergency Evacuation Plan).*
- *If public building discuss the issue and ascertain the types of issues and disabilities that your staff could be presented with. Once this has been done you can then design, agree and record the procedures (GEEP's – Generic Emergency Evacuation Plans). It would be difficult to practice a GEEP with the public, but the issue could be simulated by role play by a member of staff to test the system.*
- *Hearing impairments – How will they know alarm has gone off, flashing lights, tremblar alarms, staff actions?*
- *Vision impairment – How will they find way out, tactile signs, buddy systems, staff actions?*
- *Mobility impairment – How will they escape- horizontal and vertical? Assistance to person, specialist equipment, use of safe refuges?*
- *If person cannot escape with assistance immediately then consider use of safe refuges (fire protected) as temporary holding areas whilst you arrange for assistance, equipment as necessary to evacuate the individual (if necessary).*
- *Use of communications to ascertain type and level of risk so that a decision can be made as to the need to evacuate persons held in safe refuges.*
- *Think about other hidden disabilities such as angina and arthritis, or any other issue that could raise concerns during an evacuation e.g. heavily pregnant ladies, temporary injuries etc.*

Element 6

1. **Outline** the factors to consider when carrying out a fire risk assessment of a workplace. **(8)**

June 2000 Paper A2 Question 7 (NGC)

Reference to:

- *the probability of fire breaking out and its possible magnitude, affected by:*
  - *possible ignition sources*
  - *the quantities of flammable and combustible materials*
  - *the layout of the workplace*
- *the means of detection / raising the alarm*
  - *the siting and testing of detectors and call-points*
  - *the siting and audibility of alarms*
  - *the means of contacting the emergency services*
- *fire-fighting measures*
  - *the siting, suitability and maintenance of fire extinguishers, sprinkler systems, etc*
  - *training of personnel in the use of extinguishers*
- *evacuation*
  - *the adequacy of emergency signs*
  - *the provision and testing of emergency lighting*
  - *the number of people to be evacuated and particular groups at risk*
  - *the adequacy of escape routes*
  - *staff training in evacuation procedures*

2.   In relation to a workplace fire risk assessment, **outline** the issues that should be taken into account when accessing the *means of escape*. **(8)**

Guide to the NEBOSH Fire Safety and Risk Management Certificate, September 2005 Paper FC1 Question 11

- *How long will it take for people to escape once they are aware of fire?*
- *Is this time reasonable?*
- *Are there enough exits?*
- *Are exits in right places?*
- *Do you have suitable means of escape for all people including disabled?*
- *What fire protection is there to the escape routes and could a fire happen that would affect all escape routes?*
- *Are escape routes easily identifiable?*
- *Are exit routes free from obstructions and blockages?*
- *Are exit routes suitably lit at all times?*
- *Have staff been trained in the use of the escape routes.*

Index